ANGULAR MOMENTUM
IN QUANTUM MECHANICS

PRINCETON LANDMARKS
IN MATHEMATICS AND PHYSICS

Non-standard Analysis,
by Abraham Robinson

General Theory of Relativity,
by P.A.M Dirac

Angular Momentum in Quantum Mechanics,
by A. R. Edmonds

Mathematical Foundations of Quantum Mechanics,
by John von Neumann

Introduction to Mathematical Logic,
by Alonzo Church

Convex Analysis,
by R. Tyrrell Rockafellar

Riemannian Geometry,
by Luther Pfahler Eisenhart

The Classical Groups,
by Hermann Weyl

Topology from the Differentiable Viewpoint,
by John Milnor

ANGULAR MOMENTUM IN QUANTUM MECHANICS

BY

A. R. EDMONDS

PRINCETON UNIVERSITY PRESS
PRINCETON, NEW JERSEY

Copyright © 1957, 1960 by Princeton University Press
Copyright renewed 1985

ALL RIGHTS RESERVED

LCC 77-39796
ISBN 0-691-07912-9
ISBN 0-691-02589-4 (pbk.)

Second edition, 1960

Revised printing, 1968

Third printing, with corrections, 1974

Fourth printing, and first paperback printing, for the Princeton
Landmarks in Mathematics and Physics series, 1996

7 9 10 8 6

Printed in the United States of America

PREFACE

The concepts of angular momentum and rotational invariance play an important part in the analysis of physical systems. They have a special significance in quantum mechanics, for here we find that calculations may be divided in a natural way into two parts, namely (i) the computation of quantities which are invariant under rotations (for example the Slater integrals of atomic spectroscopy) and (ii) the evaluation of expressions which depend only on the rotational properties of the various operators and state vectors involved. It is remarkable that the structure of an expression of this latter kind is primarily a function of the complexity of the system being studied (e.g. the number of angular momenta in the coupling scheme) and is relatively independent of its precise physical nature. This fact has made it possible to develop a very general theory of angular momentum algebra, from which can be derived computational methods applicable to problems in such fields as atomic, molecular and nuclear spectroscopy, nuclear reactions, and the angular correlation of successive radiations from nuclei.

It has been my aim not only to give an account of this theory, but also to provide a practical manual for the physicist who wishes to use the associated computational methods. To this end I have paid attention to questions of notation and phase convention and have included tables of formulas and references to numerical compilations, so as to facilitate the evaluation of the various coefficients defined in the text.

The reader is assumed to have a general knowledge of quantum mechanics; an acquaintance with the theory of group representations should not be necessary.

The text is based upon the notes of lecture courses given during the last few years in the Universities of Birmingham, Manchester, Paris, Copenhagen, and Uppsala. The greater part of the writing was done while I was a member of the CERN Theoretical Study Division in Copenhagen. I am grateful to Professor Niels Bohr for the privilege of working during that time in the friendly and stimulating atmosphere of his Institute. A number of colleagues have contributed by discussions and criticisms. In particular I should like to thank K. Alder, G. Field, B. H. Flowers, P. O. Olsson and A. Winther.   London, 1957

PREFACE TO SECOND EDITION

The reprinting of this book has given me the opportunity of correcting a number of errors which appeared in the first edition, and of bringing references up to date.

I am grateful to all those people who have taken the trouble to bring errors to my notice.

London, 1960.   A. R. Edmonds

TABLE OF CONTENTS

CHAPTER 1. Group Theoretical Preliminaries   3
  1.1. Introduction   3
  1.2. Elementary Theory of Groups   5
  1.3. The Euler Angles   6
  1.4. Representation Theory   8

CHAPTER 2. The Quantization of Angular Momentum 10
  2.1. Definition of Angular Momentum in Quantum Mechanics 10
  2.2. Angular Momentum of a System of Particles 12
  2.3. Representation of the Angular Momentum Operators . . 13
  2.4. The Physical Significance of the Quantization of Angular Momentum 18
  2.5. The Eigenvectors of the Angular Momentum Operators J^2 and J_z 19
  2.6. The Spin Eigenvectors 25
  2.7. Angular Momentum Eigenfunctions in the Case of Large l 27
  2.8. Time Reversal and the Angular Momentum Operators . 29

CHAPTER 3. The Coupling of Angular Momentum Vectors . . . 31
  3.1. The Addition of Angular Momenta 31
  3.2. Commutation Relations between Components of J_1, J_2, and J 35
  3.3. Selection Rules for the Matrix Elements of J_1 and J_2 . . 35
  3.4. The Choice of the Phases of the States $w(\gamma j_1 j_2 jm)$ . . . 36
  3.5. The Vector Coupling Coefficients 37
  3.6. Computation of the Vector Coupling Coefficients 42
  3.7. The Wigner 3-j Symbol 45
  3.8. Tabulation of Formulas and Numerical Values for Vector-Coupling Coefficients 50
  3.9. Time Reversal and the Eigenvectors Resulting from Vector Coupling 51

CHAPTER 4. The Representations of Finite Rotations 53
  4.1. The Transformations of the Angular Momentum Eigenvectors under Finite Rotations 53
  4.2. The Symmetries of the $\mathfrak{D}^{(j)}_{m'm}$ 59
  4.3. Products of the $\mathfrak{D}^{(j)}_{m'm}(\alpha\beta\gamma)$ 60
  4.4. Recursion Relations for the $d^{(j)}_{m'm}(\beta)$ 61
  4.5. Computation of the $d^{(j)}_{m'm}(\beta)$ 61
  4.6. Integrals Involving the $\mathfrak{D}^{(j)}_{m'm}(\alpha\beta\gamma)$ 62
  4.7. The $\mathfrak{D}^{(j)}_{m'm}(\omega)$ as Angular Momentum Eigenfunctions . . 64
  4.8. The Symmetric Top 65

CHAPTER 5. Spherical Tensors and Tensor Operators 68
  5.1. Spherical Tensors 68
  5.2. The Tensor Operators in Quantum Mechanics 71
  5.3. Factorization of the Matrix Elements of Tensor Operators (Wigner-Eckart Theorem) 73
  5.4. The Reduced Matrix Elements of a Tensor Operator . . 75
  5.5. Hermitian Adjoint of Tensor Operators 77
  5.6. Electric Quadrupole Moment of Proton or Electron. . . 78
  5.7. The Gradient Formula 79
  5.8. Expansion of a Plane Wave in Spherical Waves 80
  5.9. Vector Spherical Harmonics 81
  5.10. Spin Spherical Harmonics 85
  5.11. Emission and Absorption of Particles 85

CHAPTER 6. The Construction of Invariants from the Vector-Coupling Coefficients 90
  6.1. The Recoupling of Three Angular Momenta 90
  6.2. The Properties of the 6-j Symbol 92
  6.3. Numerical Evaluation of the 6-j Symbol 97
  6.4. The 9-j Symbol 100

CHAPTER 7. The Evaluation of Matrix Elements in Actual Problems 109
  7.1. Matrix Elements of the Tensor Product of Two Tensor Operators 109
  7.2. Selected Examples from Atomic, Molecular and Nuclear Physics 113

APPENDIX 1. Theorems Used in Chapter 3 121

APPENDIX 2. Approximate Expressions for Vector-Coupling Coefficients and 6-j Symbols 122

Tables 1–5 124

Cited References and Bibliography 133

Index 141

ANGULAR MOMENTUM
IN QUANTUM MECHANICS

CHAPTER 1

Group Theoretical Preliminaries

1.1. Introduction

The subject of this book is the detailed development of the uses of the principle of conservation of angular momentum in the analysis of physical systems. While this principle is by no means trivial in classical mechanics, it is of fundamental importance in the quantum mechanics of many-particle systems. Such systems include the more complex atoms, the atomic nuclei treated from the point of view of the independent particle model, and experiments in which particles are emitted from or absorbed by nuclei.

We shall first discuss the relevance of conservation of the angular momentum of a system in classical mechanics, and see how it is related to the symmetry of the Hamiltonian of the system with respect to rotations of the frame of reference. Thus even in a classical analysis we find that the theory of the group of rotations in three dimensions is bound up with the idea of angular momentum.

THE SYMMETRY OF THE HAMILTONIAN.[1] A constant of the motion is a function of the canonical variables which does not change with time, and in the classical mechanics a knowledge of all the constants of the motion of a system amounts to a solution of the equations of motion. Now for any function u of the canonical variables which does not depend explicitly on the time the Poisson bracket of the function with the Hamiltonian is zero; for

$$\frac{du}{dt} = [u, H] = 0.$$

An *infinitesimal contact transformation* may be defined as a contact transformation which changes the canonical variables q_i, p_i ($i = i, 2, \ldots, n$) by an infinitesimal amount:

$$q_i \to q_i' = q_i + \delta q_i$$

$$p_i \to p_i' = p_i + \delta p_i$$

The generating function F of the infinitesimal transformation differs

---

[1] For a more detailed treatment see any advanced textbook on classical mechanics, e.g. Goldstein (1950).

only infinitesimally from the generating function of the identity transformation, which is $\sum_i q_i p'_i$. We may write it therefore as

$$F = \sum_i q_i p'_i + \varepsilon\, G(q, p')$$

where ε is an infinitesimal parameter. It is customary to call $G(q, p')$ the generating function of the infinitesimal transformation, in spite of the fact that this is also the name of the quantity F. It may be shown that the change in a function u of the canonical variables due to this transformation is

$$\delta u = \varepsilon [u, G].$$

Hence replacing u by the Hamiltonian H, we have

$$\delta H = \varepsilon [H, G].$$

Thus we deduce that the constants of the motion are the generating functions of those infinitesimal contact transformations which leave the Hamiltonian invariant.

We find in particular that the angular momentum components are the generating functions of the infinitesimal rotations about the corresponding axes of the frame of reference. Thus if the angular momentum is a constant of the motion, then the Hamiltonian of the system is symmetric with respect to rotation of the frame of reference about the origin. We say that the *group of the Hamiltonian*, i.e. the group of transformations which leave the Hamiltonian invariant, contains the group $SO(3)$ of rotations in three-dimensional space. This fact is of importance in quantum mechanics, for the theorem of Wigner-Eckart states[2] that if T is an element of the group G_H of the Hamiltonian H, and if u is an eigenvector of H, then Tu is also an eigenvector of H with the same eigenvalue. This implies that all eigenvectors of H belonging to a given irreducible representation of G_H have the same eigenvalue, i.e. are degenerate in energy; however this statement contains group theoretical terminology which has not yet been explained.

In the case of a system with rotational symmetry, the theorem implies that, as is well known, the angular momentum eigenvectors are eigenvectors of the energy and that the set of states with the same total angular momentum and different values of the z-component is degenerate.

---

[2] See Wigner (1927), Eckart (1930).

1.2. Elementary Theory of Groups[3]

The concept of group is a generalization of the properties of a large number of systems of mathematical interest; such systems as the set of all permutations of n objects, the set of all rotations of a rigid body, the set of all nonsingular linear transformations on a given vector space.

An *abstract group* is defined without reference to any particular physical or mathematical system. It is in fact a set of elements among which a law of composition is defined such that the composition of any two elements a and b of the group taken in this order and denoted by ba is an element of the set.[4]

We must add to this property the following conditions:

1. The associative law $c(ba) = (cb)a$.
2. There exists a unit element 1, which leaves any element a unaltered on composition with it:

$$1a = a1 = a.$$

3. To each element a corresponds an inverse a^{-1} which gives on composition with a the unit element:

$$aa^{-1} = a^{-1}a = 1.$$

The number of elements in a group, its *order*, may be finite, or denumerably or nondenumerably infinite. Among finite groups are the symmetry groups of the regular solids and the permutation groups on a finite number of objects. The positive and negative integers form a group of denumerably infinite order with respect to addition. The simplest group with a nondenumerable set of elements is the set of real numbers with respect to addition, or equivalently the set of all translations of a point on a line.

A *subgroup* h of a group g is a set of elements of g which itself fulfills the group conditions. The unit element must thus belong to h, and if a and b both belong to h, then so do a^{-1} and ba.

The groups we shall be concerned with are those with a nondenumerable infinity of elements. Let us consider first the set of all nonsingular linear homogeneous transformations on an n-dimensional vector space; we suppose the transformation matrices to have complex coefficients. This set clearly forms a group with respect to composition of the transformations (i.e. to matrix multiplication); it is known as the full linear group $GL(n)$. Restriction of these transformations to unitary trans-

---

[3] The reader is referred for a more detailed treatment of the applications of the theory of groups to quantum mechanics to the well-known works of Weyl (1931), Wigner (1931), Eckart (1930), van der Waerden (1931), and Bauer (1933).
[4] Note that in general $ab \neq ba$.

formations gives us the *unitary* group $U(n)$, which is a subgroup of $GL(n)$, this relation being symbolized by

$$U(n) \subset GL(n).$$

We may make the further restriction that the unitary matrices have determinant $+1$, i.e. are *unimodular*. The resulting group is called the *special* unitary group $SU(n)$. The group of all real linear homogeneous transformations on an n-dimensional space which preserve the distance between two points, defined in the Euclidean sense, i.e. the rotations and reflections about the origin, is called the orthogonal group $O(n)$. It corresponds to the set of all real $n \times n$ orthogonal matrices. We shall be concerned particularly with rotations in 3-space, namely with the unimodular orthogonal group $SO(3)$.

We come now to the question of how to label the elements of a group of the type with which we have just been dealing. Evidently in the case of the elements of $GL(n)$ we would need n^2 complex numbers to specify any element, since the matrix elements are independent. Imposition of restrictions (e.g. orthogonality) on the matrix elements will reduce the number of independent quantities; and in the case of the rotation group $SO(3)$ we need only three real numbers, a fact well known from geometry. For any rotation of a rigid body may be symbolized by three real numbers.

1.3. The Euler Angles

The most useful way of defining these three numbers, i.e. of parameterizing the rotation group, is that of Euler; there are, however, several conventions in existence for choosing the so-called Euler angles. We shall consider this choice with some care, for ambiguities in the definition of the Euler angles entail confusion in questions of the phases of matrix elements of finite rotations, etc.

The convention we shall use is that employed frequently by workers in the theories of molecular spectra (Herzberg 1939) and of the collective model of the atomic nucleus (Bohr 1952). It differs, for example, from those of Wigner (1931) (who employs a left-handed frame of reference) and of Casimir (1931).

The general displacement of a rigid body due to a rotation about a fixed point may be obtained by performing three rotations about two of three mutually perpendicular axes fixed in the body. We shall assume a right-handed frame of axes; we shall further define a *positive* rotation about a given axis to be one which would carry a right-handed screw in the positive direction along that axis. Thus a rotation about the z-axis which carried the x-axis into the original position of the y-axis

1.3 · THE EULER ANGLES

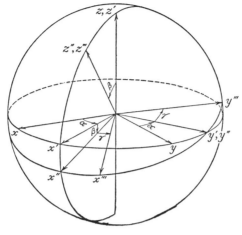

Fig. 1.1

would be considered to be positive. The rotations (see Fig. 1.1) are to be performed successively in the order:

1) A rotation $\alpha(0 \leq \alpha < 2\pi)$ about the z-axis, bringing the frame of axes from the initial position S into the position S'. The axis of this rotation is commonly called the *vertical*.
2) A rotation $\beta(0 \leq \beta < \pi)$ about the y-axis of the frame S', called the *line of nodes*. Note that its position is in general different from the initial position of the y-axis of the frame S. The resulting position of the frame of axes is symbolized by S''.
3) A rotation $\gamma(0 \leq \gamma < 2\pi)$ about the z-axis of the frame of axes S'', called the *figure axis*; the position of this axis depends on the previous rotations α and β. The final position of the frame is symbolized by S'''. Although the possible values of α, β and γ are restricted, we do not have a 1 : 1 correspondence between rotations and parameters for all possible rotations; for example a rotation symbolized by $(\alpha 0 \gamma)$ is identical with that symbolized by $(\alpha' 0 \gamma')$ if $\alpha + \gamma = \alpha' + \gamma'$.

It should be noted that the polar coordinates φ, θ with respect to the original frame S of the z-axis in its final position are identical with the Euler angles α, β respectively.

In the description of the general rotation just given, the rotations β and γ have been defined with respect to the frame of reference carried with the moving body. It is convenient in many applications always to refer rotations to the original fixed frame of axes S.

We represent a rotation θ about an axis ξ by the operator $D_\xi(\theta)$. Operators corresponding to successive rotations are ordered from right to left.

Thus the rotation 1) above is written as $D_z(\alpha)$ and the composition of the three successive rotations 1), 2) and 3) as $D_{z''}(\gamma)\, D_{y'}(\beta)\, D_z(\alpha)$.

Now the rotation $D_{y'}(\beta)$ about the axis y' in S' is equivalent to the composition $D_z(\alpha)\, D_y(\beta)\, D_z(-\alpha)$ of three successive rotations about axes in S.

Similarly $D_{z''}(\gamma)$ is equivalent to $D_{y'}(\beta)\, D_{z'}(\gamma)\, D_{y'}(-\beta)$ and hence to $D_z(\alpha)\, D_y(\beta)\, D_z(\gamma)\, D_y(-\beta)\, D_z(-\alpha)$.

It follows that

(1.3.1) $$D_{z''}(\gamma)\, D_{y'}(\beta)\, D_z(\alpha) \equiv D_z(\alpha)\, D_y(\beta)\, D_z(\gamma).$$

I.e. the rotation defined by 1), 2) and 3) is equivalent to the rotation which is the result of carrying out in order the following rotations of the body about the fixed axes of S:

1') a rotation γ about the z axis
2') a rotation β about the y axis
3') a rotation α about the z axis.

1.4. Representation Theory

A very important part, from a physical point of view, of the theory of groups is that concerned with the *representation* of the elements of a group by linear transformations.

We mean by a *representation of degree n* of a group G that to every element a of G is assigned a linear transformation $T(a)$ on a vector space $\mathfrak{R}_n$ of dimension n in such a way that these linear transformations obey the law of composition:

(1.4.1) $$T(a) \cdot T(b) = T(ab).$$

It may be the case that to each group element corresponds a distinct transformation; we speak then of a *faithful* representation. On the other hand we get a representation which fulfills (1.4.1) by choosing for each and every transformation the identity transformation.

When a definite coordinate system is chosen in the space $\mathfrak{R}_n$ each transformation $T(a)$ corresponds to a square nonsingular matrix. The orthogonal unit vectors which establish this coordinate system are called the *basis* of the representation. If we replace the coordinate system by another obtained from it by a transformation S, the group element a will be represented by the transformation $ST(a)S^{-1}$. We have again a representation of G, which is said to be *equivalent* to the former one.

Let us consider how such a representation might arise; we take for example the set of all sufficiently well-behaved functions on the surface of a sphere. A given function may be represented by a vector in a function space whose basis vectors are chosen functions forming a complete orthogonal set—say, the spherical harmonics. A rotation of the sphere will induce linear transformations in this function space; these give a representation of the rotation group.

REDUCIBILITY. Suppose there exists a subspace $\mathcal{R}'$ of $\mathcal{R}$ such that all vectors lying in this subspace are transformed by a given transformation T into vectors of $\mathcal{R}'$. We say then that the subspace $\mathcal{R}'$ is *invariant* under the transformation T. If $\mathcal{R}'$ is invariant under all transformations $T(a)$ representing the group G, the transformations $T'(a)$ which are induced in $\mathcal{R}'$ themselves give a representation of G. If we picture the transformations as matrices, then we may choose such a basis that all the representation matrices in a given representation take the form of Fig. 1.2, where the submatrix P corresponds to the

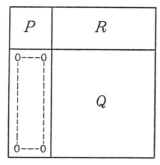

Fig. 1.2

transformations on the subspace $\mathcal{R}'$. (The rectangular submatrix R will usually, but not necessarily, contain nothing but zeros.) A representation based on a space $\mathcal{R}$ is called *irreducible* if $\mathcal{R}$ contains no subspace other than itself and the null space which is invariant under the transformations $T(a)$ representing the group G.

An example of an irreducible representation is that given by the spherical harmonics of a given order l. It is well known that (due to the invariance of the Laplace equation under rotations of the frame of reference) a spherical harmonic Y_{lm} is transformed by rotation of the frame of reference into a function expressible as a sum of spherical harmonics with the same l but with m running over the whole range $-l \leq m \leq l$, each with an appropriate coefficient; the coefficients are the matrix elements of the representation. Since any Y_{lm} may be transformed by some rotation into a function containing any other with the same l, the representation of degree $2l + 1$ whose basis is the set of functions $Y_{l,-l}, Y_{l,-l+1}, \cdots, Y_{l,l-1}, Y_{l,l}$ is irreducible.

CHAPTER 2

The Quantization of Angular Momentum

2.1. Definition of Angular Momentum in Quantum Mechanics

ANGULAR MOMENTUM IN CLASSICAL MECHANICS. In the classical theory the angular momentum of a system of n massive particles is defined as a vector, given by

$$\mathbf{L} = \sum_{i=1}^{n} \mathbf{r}_i \times \mathbf{p}_i$$

where $\mathbf{r}_i$, $\mathbf{p}_i$ are the position vector and linear momentum respectively of the ith particle. We may write down a similar integral expression for a continuous distribution of matter. Provided that there are no external torques operating on the system, all three components of $\mathbf{L}$ are constants of the motion, and may take any finite values whatever.

THE INTRODUCTION OF QUANTIZATION. The historic paper of Bohr (1913) on the spectrum of the hydrogen atom introduced for the first time the postulate that the angular momentum of a system was quantized, i.e. that it could only take values which were integer multiples of the quantum of action h times $1/2\pi$. Sommerfeld (1916) suggested that the direction as well as the magnitude of the angular momentum of an electron in a closed orbit was quantized; that is, that only certain directions of orientation of the angular momentum vector with respect to a fixed axis were possible.

From that time onwards spectroscopists studying the structure of atoms made use of empirical rules for dealing with the coupling of the angular momenta involved (cf. Landé (1923)). Difficulties in interpretation of these rules continued until the discovery of wave and matrix mechanics, and the establishment of a definite procedure for making the step from the classical to the quantum theory.

DERIVATION OF THE COMMUTATION RULES. In classical mechanics the angular momentum of a particle about a point O is defined as

(2.1.1) $$\mathbf{L} = \mathbf{r} \times \mathbf{p}$$

where $\mathbf{r}$ is the position vector of the particle with respect to O and $\mathbf{p}$ is its linear momentum.

In quantum mechanics the components of position and linear momentum of a particle obey the commutation relations[1]

---
[1] We employ the alternative and equivalent notations x_1, x_2, x_3 or x, y, z for components of positions, etc.

2.1 · DEFINITION OF ANGULAR MOMENTUM

$$[x_i, p_j] = i\hbar \delta_{ij}; \quad [x_i, x_j] = 0; \quad [p_i, p_j] = 0$$

where $\quad i, j = 1, 2, 3$

We apply these relations to find the commutation rules for the components of angular momentum. For example

$$[L_x, L_y] = (yp_z - zp_y)(zp_x - xp_z) - (zp_x - xp_z)(yp_z - zp_y)$$
$$= yp_x(p_z z - zp_z) + xp_y(zp_z - p_z z) = i\hbar(xp_y - yp_x)$$

Thus we obtain

(2.1.2) $\quad [L_x, L_y] = i\hbar L_z; \quad [L_y, L_z] = i\hbar L_x; \quad [L_z, L_x] = i\hbar L_y.$

DIFFERENTIAL OPERATOR EXPRESSIONS FOR THE COMPONENTS OF ANGULAR MOMENTUM. We may express the operators of angular momentum in the differential operator form; we take $p_x = -i\hbar(\partial/\partial x)$ etc.[2]

(2.1.3)
$$L_x = -i\hbar \left(y \frac{\partial}{\partial z} - z \frac{\partial}{\partial y} \right)$$
$$L_y = -i\hbar \left(z \frac{\partial}{\partial x} - x \frac{\partial}{\partial z} \right)$$
$$L_z = -i\hbar \left(x \frac{\partial}{\partial y} - y \frac{\partial}{\partial x} \right)$$

These equations may be written in terms of the spherical polar coordinates

(2.1.4)
$$L_x = i\hbar \left(\sin \varphi \frac{\partial}{\partial \theta} + \cot \theta \cos \varphi \frac{\partial}{\partial \varphi} \right)$$
$$L_y = i\hbar \left(-\cos \varphi \frac{\partial}{\partial \theta} + \cot \theta \sin \varphi \frac{\partial}{\partial \varphi} \right)$$
$$L_z = -i\hbar \frac{\partial}{\partial \varphi}$$

Thus we see that the angular momentum operators are proportional to the operators of infinitesimal rotations[3] (cf. Dirac's displacement operators, Dirac (1947) §25).

The square of the total angular momentum is defined as

(2.1.5) $\quad \mathbf{L}^2 = L_x^2 + L_y^2 + L_z^2$

This operator commutes with L_x, L_y, and L_z as may be shown by use of (2.1.2). It is given in terms of the spherical differential operators by

---
[2] Schiff (1949) p. 20.
[3] Goldstein (1950) pp. 124, 263.

(2.1.6) $$\mathbf{L}^2 = -\hbar^2 \left[\frac{1}{\sin\theta} \frac{\partial}{\partial\theta} \left(\sin\theta \frac{\partial}{\partial\theta} \right) + \frac{1}{\sin^2\theta} \frac{\partial^2}{\partial\varphi^2} \right]$$

2.2. Angular Momentum of a System of Particles

PRELIMINARY REMARKS. In classical mechanics the angular momentum of a system of n particles relative to a point O is given by

(2.2.1) $$\mathbf{L} = \sum_{i=1}^{n} \mathbf{r}_i \times \mathbf{p}_i = \sum_{i=1}^{n} \mathbf{L}_i$$

where $\mathbf{r}_i$, $\mathbf{p}_i$, and $\mathbf{L}_i$ are the position vector with respect to O, the linear momentum, and angular momentum respectively of the ith particle. Since the quantum mechanical operators relating to different particles commute, we may take over this definition into quantum mechanics with the knowledge that the components L_x, L_y, L_z of $\mathbf{L}$ obey the same commutation rules as the components of the angular momenta $\mathbf{L}_i$ of individual particles.

Now we may write down differential operator expressions for the components of the total angular momentum in terms of the $3n$ coordinates of the particles in the obvious way, namely by writing down an expression corresponding to (2.1.3) or (2.1.4) for each of the n particles. However it is instructive to go about the problem in a different way.[4]

THE INVARIANTS AND EULER ANGLES OF A SYSTEM OF n PARTICLES. A number of invariants, i.e. quantities whose values are unchanged by rotation of the frame of coordinates, may be built up from the $3n$ coordinates of the n particles. There are obviously the n lengths r_i of the position vectors $\mathbf{r}_i$. There are also the *scalar products* of the vectors taken two at a time. We must decide how many of these scalar products need to be specified to fix the relative orientation of all the vectors. If we choose any two vectors, say $\mathbf{r}_1$ and $\mathbf{r}_2$, and specify the scalar product $(\mathbf{r}_1 \cdot \mathbf{r}_2)$, each of the remaining $n - 2$ vectors is fixed by specifying the values of two scalar products. There are thus $1 + 2(n - 2) = 2n - 3$ independent scalar products. The total number of independent invariants is thus $3n - 3$. There remain 3 independent quantities; these may be supposed to determine the 3 Euler angles (see (1.3)) of a moving frame of reference which is associated with the motion of the n particles. If the n particles move together rigidly about the origin of coordinates, i.e. if the $3n - 3$ invariants are constants of the motion we have the well-known case of the rotation of a rigid body; the moving frame is fixed in the body. If the motion is not rigid and $n > 2$, it is not so easy to specify the Euler angles in terms of the coordinates; nevertheless there is no real ambiguity in their specification since, as we have seen, there are only 3 independent quantities with which they may be associated.

---

[4] Cf. Sommerfeld (1939) Vol. II p. 776.

THE TOTAL ANGULAR MOMENTUM IN TERMS OF THE EULER ANGLES. The operators of infinitesimal rotations about the instantaneous Euler axes (the vertical, the line of nodes, and the figure axis) are first expressed in terms of the infinitesimal rotations about the fixed x, y, and z axes. It is well known that these infinitesimal rotations may be compounded as vectors; it follows that the operator of infinitesimal rotation about the line of nodes is given by

$$\frac{\partial}{\partial \beta} = -\sin \alpha \frac{\partial}{\partial \alpha_x} + \cos \alpha \frac{\partial}{\partial \alpha_y}$$

where α_x and α_y are angles, analogous to α, measured about the fixed x and y axes respectively. Similarly the infinitesimal rotation about the figure axis is

$$\frac{\partial}{\partial \gamma} = \cos \alpha \sin \beta \frac{\partial}{\partial \alpha_x} + \sin \alpha \sin \beta \frac{\partial}{\partial \alpha_y} + \cos \beta \frac{\partial}{\partial \alpha}$$

We have in analogy with $L_z = -i\hbar(\partial/\partial\alpha)$ also

$$L_x = -i\hbar \frac{\partial}{\partial \alpha_x}, \qquad L_y = -i\hbar \frac{\partial}{\partial \alpha_y}$$

and we may invert the equations for $\partial/\partial\alpha$, $\partial/\partial\beta$, $\partial/\partial\gamma$ to obtain

(2.2.2)
$$L_x = -i\hbar \left\{ -\cos \alpha \cot \beta \frac{\partial}{\partial \alpha} - \sin \alpha \frac{\partial}{\partial \beta} + \frac{\cos \alpha}{\sin \beta} \frac{\partial}{\partial \gamma} \right\}$$

$$L_y = -i\hbar \left\{ -\sin \alpha \cot \beta \frac{\partial}{\partial \alpha} + \cos \alpha \frac{\partial}{\partial \beta} + \frac{\sin \alpha}{\sin \beta} \frac{\partial}{\partial \gamma} \right\}$$

$$L_z = -i\hbar \frac{\partial}{\partial \alpha}$$

The square of the total angular momentum is given by (2.1.5) and (2.2.2) as

(2.2.3)
$$\mathbf{L}^2 = \hbar^2 \left\{ -\frac{\partial^2}{\partial \beta^2} - \cot \beta \frac{\partial}{\partial \beta} \right.$$

$$\left. - \frac{1}{\sin^2 \beta} \left(\frac{\partial^2}{\partial \alpha^2} + \frac{\partial^2}{\partial \gamma^2} \right) + \frac{2 \cos \beta}{\sin^2 \beta} \frac{\partial^2}{\partial \alpha\, \partial \gamma} \right\}$$

2.3. Representation of the Angular Momentum Operators

EXTENDED DEFINITION OF THE OPERATORS. In this section we derive the Hermitian matrix representations of the system of operators (2.1.2). For the purpose of this discussion the operators are supposed to be *defined* by the commutation relations (2.1.2); we shall see that this gives a greater content to our theory than just assuming that all their properties are expressed in terms of differential operators. In particular, this definition permits the existence of spin, which, as is well known, cannot exist in the framework of classical mechanics. To emphasize the extension of this definition, we shall employ the symbols J_x, J_y, and J_z for the components of angular momentum in general. L will be kept to symbolize orbital angular momentum only. It is convenient to use the non-Hermitian operators J_+ and J_-, defined by

(2.3.1) $$J_+ = J_x + iJ_y; \quad J_- = J_x - iJ_y$$

They obey the commutation relations

(2.3.2) $$[\mathbf{J}^2, J_\pm] = 0; \quad [J_z, J_+] = \hbar J_+$$
$$[J_z, J_-] = -\hbar J_-; \quad [J_+, J_-] = 2\hbar J_z.$$

BASIS OF THE REPRESENTATION. We choose as basis of our representation the simultaneous normalized eigenvectors of the commuting operators $\mathbf{J}^2$ and J_z. The choice of J_z is quite arbitrary. The eigenvectors $u(jm)$ are labeled by the symbols j and m which are related by a 1 : 1 correspondence to the eigenvalues of $\mathbf{J}^2$ and J_z. Eigenvectors with distinct symbols j and/or m are supposed to have distinct eigenvalues λ_j and/or λ_m of $\mathbf{J}^2$ and J_z and vice versa. That is, eigenvectors $u(jm)$ with different values of j and/or m are orthogonal.[5] All the elements of the basis which we consider are supposed to possess the same set of eigenvalues with respect to those operators Γ which, together with $\mathbf{J}^2$ and J_z, form a complete commuting set for the system being studied (cf. Dirac (1947) p. 57).

The matrix element of any operator Θ is defined in the angular momentum representation by the relation[6]

$$\Theta u(\gamma \; j \; m) = \sum u(\gamma' \; j' \; m')(\gamma' \; j' \; m'|\Theta|\gamma \; j \; m)$$
$$(\gamma' \; j' \; m'|\Theta|\gamma \; j \; m) = \Big(u(\gamma' \; j' \; m'), \Theta u(\gamma \; j \; m)\Big)$$

where the expression on the right is the Hermitian or scalar product; the assumed orthonormality of the $u(jm)$ implies that

(2.3.3) $$\Big(u(j' \; m'), u(j \; m)\Big) = \delta_{j'j}\delta_{m'm}$$

The correspondence between the symbols m and the eigenvalues λ_m of J_z is made by writing

(2.3.4) $$J_z u(j \; m) = m\hbar u(j \; m)$$

---

[5]See Dirac (1947) p. 32.
[6]γ symbolizes the eigenvalues of the operators Γ mentioned above; it will be omitted where not relevant.

2.3 · REPRESENTATION OF THE OPERATORS

Since all the angular momentum operators commute with $\mathbf{J}^2$ they send any $u(j\,m)$ into another vector which is also an eigenvector of $\mathbf{J}^2$ with the same eigenvalue (i.e. the same j). For we have

(2.3.5) $\qquad \mathbf{J}^2 J u(j\,m) = J \mathbf{J}^2 u(j\,m) = \lambda_j J u(j\,m)$

where J is any of the J_x, J_y, J_z or a linear combination of them, and λ_j is the eigenvalue of $\mathbf{J}^2$ corresponding to the symbol j. We may therefore restrict our considerations to a subset of eigenvectors $u(j\,m)$ which all have the same eigenvalue of $\mathbf{J}^2$, i.e. which are all labeled by the same j.

Let us consider the matrix component of the equation

$$J_z J_+ - J_+ J_z = \hbar J_+$$

(see (2.3.2)) between $j\,m'$ and $j\,m$. We have

(2.3.6) $\qquad (m'-m)\hbar(j\,m'|J_+|j\,m) = \hbar(j\,m'|J_+|j\,m)$

I.e. the only nonvanishing matrix elements of J_+ are for $m'-m=1$. Hence we may write

(2.3.7) $\qquad J_+ u(j\,m) = x_m \hbar u(j\,m+1)$

where x_m is a number which may be complex. A similar argument shows that

(2.3.8) $\qquad J_- u(j\,m) = x'_m \hbar u(j\,m-1)$

It is easy to see from the definition (2.3.1) of J_+ and J_- and from the fact that the operators J_x and J_y are Hermitian that x_m and x'_{m+1} are complex conjugate,

(2.3.9) $\qquad\qquad x'_{m+1} = x_m^*.$

Hence the commutation relation $J_+ J_- - J_- J_+ = 2\hbar J_z$ implies that $x_{m-1} x_{m-1}^* - x_m^* x_m = 2m$. I.e. we have a difference equation for $|x_m|^2$:

(2.3.10) $\qquad\qquad |x_{m-1}|^2 - |x_m|^2 = 2m$

The general solution contains an arbitrary constant C:

(2.3.11) $\qquad\qquad |x_m|^2 = C - m(m+1)$

Now for any finite value of C the right-hand side becomes negative for sufficiently large positive or negative values of m; however $|x_m|^2$ is necessarily non-negative. The apparent contradiction is removed when we see that the relation (2.3.6) is satisfied for *any* values of m when the matrix element of J_+ is zero, i.e. when $|x_m|^2$ is zero. We may therefore suppose that $|x_m|^2$ takes nonzero values only over a restricted range of values of m:

(2.3.12) $\qquad m = \underline{m}+1,\, \underline{m}+2,\, \cdots,\, \bar{m}-2,\, \bar{m}-1$

where the lower and upper bounds $\underline{m}$ and $\bar{m}$ differ by an integer. The eigenvectors $u(j\,m)$ which enter into the representation thus have m values $\underline{m}+1, \underline{m}+2, \ldots, \bar{m}-1, \bar{m}$. The bounding values $\underline{m}$ and $\bar{m}$ are found by solving the quadratic equation derived from (2.3.11):

$$0 = C - m(m+1)$$

We obtain $\underline{m} = -\tfrac{1}{2} - \tfrac{1}{2}(1+4C)^{\frac{1}{2}}$, $\bar{m} = -\tfrac{1}{2} + \tfrac{1}{2}(1+4C)^{\frac{1}{2}}$

I.e.

(2.3.13) $\qquad C = \bar{m}(\bar{m}+1)$ and $\underline{m} = -\bar{m}-1$.

Since $\bar{m}$ and $\underline{m}$ differ by an integer, $2\bar{m}$ is a positive integer and $\bar{m}$ may only take the values $0, \tfrac{1}{2}, 1, \tfrac{3}{2}, 2, \ldots$.

THE EIGENVALUES OF $\mathbf{J}^2$. The operator $\mathbf{J}^2$ is given in terms of J_+ and J_- by

(2.3.14) $\qquad \mathbf{J}^2 = \tfrac{1}{2}(J_+J_- + J_-J_+) + J_z^2$.

Its eigenvalues in the scheme just considered may therefore be computed by use of the results already obtained. We must thus find λ_j in

$$\mathbf{J}^2 u(j\,m) = \lambda_j u(j\,m)$$

where m takes one of the values (2.3.12) and we make use of (2.3.4), (2.3.7), (2.3.8), (2.3.9), (2.3.11), and (2.3.13).
We get

$$\lambda_j = \frac{\hbar^2}{2}(|x_{m-1}|^2 + |x_m|^2) + m^2\hbar^2$$

$$= \frac{\hbar^2}{2}[\bar{m}(\bar{m}+1) - (m-1)m + \bar{m}(\bar{m}+1) - m(m+1)] + m^2\hbar^2$$

$$= \bar{m}(\bar{m}+1)\hbar^2$$

which is, as expected, independent of m. Now we identify $\bar{m}$ with the symbol j used to label eigenvectors of $\mathbf{J}^2$; the task of constructing the representations of the angular momentum operators is now completed.

The results are as follows:

A basis of a representation of the angular momentum operators is given by the simultaneous eigenvectors $u(j\,m)$ of $\mathbf{J}^2$ and J_z where j and m are given by

(2.3.15) $\qquad \mathbf{J}^2 u(j\,m) = \hbar^2 j(j+1) u(j\,m)$

and

$$J_z u(j\,m) = \hbar m u(j\,m)$$

and the values of j and m are subject to certain restrictions.

2.3 · REPRESENTATION OF THE OPERATORS

(i) For a given representation j is fixed and may take one of the values $0, \frac{1}{2}, 1, \frac{3}{2}, 2, \ldots$.

(ii) There are $2j + 1$ values of m allowed for a particular j, namely $m = -j, -j + 1, \ldots, j - 1, j$.

We symbolize the $(2j + 1)$-dimensional representation whose basis is given by the eigenvectors $u(j, -j), u(j, -j + 1), \ldots, u(j\,j)$ by $\mathfrak{D}^{(j)}$. It is clear that, by successive use of the operators J_+ or J_-, we may transform any vector in this set into any other. The representation is therefore in the group theoretical sense *irreducible*.

THE MATRICES OF THE ANGULAR MOMENTUM OPERATORS. The matrix elements of J_+ and J_- are given by (2.3.7), (2.3.8), (2.3.9), (2.3.11), and (2.3.13), but only up to a phase. The choice of this phase is quite arbitrary but must be followed consistently. The convention established by Condon and Shortley (1935) of taking this phase as $+1$ is now almost universal. We make this choice and obtain

(2.3.16) $\quad J_+ u(j\,m) = \hbar[(j - m)(j + m + 1)]^{\frac{1}{2}} u(j\,m+1)$

(2.3.17) $\quad J_- u(j\,m) = \hbar[(j + m)(j - m + 1)]^{\frac{1}{2}} u(j\,m-1)$

Hence the nonzero matrix elements of J_x and J_y are

(2.3.18)
$$(j\,m+1|J_x|j\,m) = \tfrac{1}{2}\hbar[(j - m)(j + m + 1)]^{\frac{1}{2}}$$
$$(j\,m-1|J_x|j\,m) = \tfrac{1}{2}\hbar[(j + m)(j - m + 1)]^{\frac{1}{2}}$$
$$(j\,m+1|J_y|j\,m) = -\tfrac{i}{2}\hbar[(j - m)(j + m + 1)]^{\frac{1}{2}}$$
$$(j\,m-1|J_y|j\,m) = \tfrac{i}{2}\hbar[(j + m)(j - m + 1)]^{\frac{1}{2}}$$

THE SPIN REPRESENTATION. The matrices of the angular momentum operators are of particular interest for the case $j = \frac{1}{2}$; they are namely

(2.3.19)

$(\tfrac{1}{2} m'\|J_x\|\tfrac{1}{2} m)$			$(\tfrac{1}{2} m'\|J_y\|\tfrac{1}{2} m)$			$(\tfrac{1}{2} m'\|J_z\|\tfrac{1}{2} m)$		
$m' \backslash m$	$+\tfrac{1}{2}$	$-\tfrac{1}{2}$	$m' \backslash m$	$+\tfrac{1}{2}$	$-\tfrac{1}{2}$	$m' \backslash m$	$+\tfrac{1}{2}$	$-\tfrac{1}{2}$
$+\tfrac{1}{2}$	0	$\hbar/2$	$+\tfrac{1}{2}$	0	$-i\hbar/2$	$+\tfrac{1}{2}$	$\hbar/2$	0
$-\tfrac{1}{2}$	$\hbar/2$	0	$-\tfrac{1}{2}$	$i\hbar/2$	0	$-\tfrac{1}{2}$	0	$-\hbar/2$

These are the *Pauli spin matrices*,[7] and are frequently written $(\hbar/2)\sigma_x$,

---

[7] W. Pauli (1927).

$(\hbar/2)\sigma_y$ and $(\hbar/2)\sigma_z$ where

(2.3.20) $$\sigma_x = \begin{pmatrix} 0 & 1 \\ 1 & 0 \end{pmatrix}; \quad \sigma_y = \begin{pmatrix} 0 & -i \\ i & 0 \end{pmatrix}; \quad \sigma_z = \begin{pmatrix} 1 & 0 \\ 0 & -1 \end{pmatrix}.$$

2.4. The Physical Significance of the Quantization of Angular Momentum

The most immediate consequence of quantization upon the angular momentum of a system is that the components no longer commute. The uncertainty principle therefore makes it impossible to measure simultaneously the values of all three (or even two) components of angular momentum.

We have the rule for the minimum uncertainties of measurement ΔA and ΔB of any two noncommuting operators A and B:

(2.4.1) $$\overline{(\Delta A)^2} \cdot \overline{(\Delta B)^2} \geq \overline{\left\{\frac{i}{2}[A,B]\right\}^2}$$

where the bars imply expectation values. Now if, as we have already supposed, we choose to measure the component along the z-axis, obtaining a value $\hbar m$, then we have for the minimum uncertainties of J_x and J_y,

(2.4.2) $$\overline{(\Delta J_x)^2} \cdot \overline{(\Delta J_y)^2} \geq \frac{\hbar^2}{4}\overline{J_z^2} = \frac{m^2 \hbar^2}{4}$$

Another striking feature of the quantization is the fact that the measured values of total angular momentum and of its component in a given direction can take only certain values, namely $\hbar^2 j(j+1)$ and $\hbar m$ ($m = -j, -j+1, \ldots, j$), thus justifying the postulates of the older theories. In fact, the expression $\hbar^2 j(j+1)$ for the square of the length of the angular momentum vector was discovered empirically by spectroscopists.[8] It is important to note that the angular momentum vector can never point exactly in the direction of the z-axis; the maximum value of m is j while the length of the vector is $\hbar\sqrt{j(j+1)}$. This is associated with the uncertainty in measurement of the x and y components. Another important feature is the possibility of half-odd integer values of the eigenvalues of angular momentum which arises, as already mentioned, from the generalization of the concept of angular momentum. We may obtain a more accurate measure of the uncertainty in J_x and J_y than the above inequality in the following way. We have

$$\overline{(\Delta J_x)^2} = \overline{(J_x - \overline{J_x})^2} = \overline{J_x^2} - (\overline{J_x})^2 = \overline{J_x^2}$$

---

[8] Cf. Landé (1923).

and a similar expression for J_y. The expectation value of the square of the angular momentum is given by

$$\overline{J^2} = \hbar^2 j(j+1) = \overline{J_x^2} + \overline{J_y^2} + \overline{J_z^2} = \overline{(\Delta J_x)^2} + \overline{(\Delta J_y)^2} + m^2\hbar^2$$

Hence

(2.4.3) $$\overline{(\Delta J_x)^2} + \overline{(\Delta J_y)^2} = \hbar^2(j^2 + j - m^2)$$

The minimum fluctuation in the measurements of J_x and J_y clearly occurs for $|m| = j$, i.e. when the angular momentum vector points as nearly as possible along the z-axis. We might imagine the vector moving in an unobservable way about the z-axis, keeping the angle between itself and the axis constant. This picture will be made more concrete when we examine the orbital wave functions, which describe the probability density of a moving particle. There is however one case in which the components J_x and J_y are sharply defined; namely when the total angular momentum is zero.

2.5. The Eigenvectors of the Angular Momentum Operators J^2 and J_z

THE EIGENFUNCTIONS OF ORBITAL ANGULAR MOMENTUM. We shall consider first the eigenvectors of J^2 and J_z when they appear in the form L^2 (2.1.6) and L_z (2.1.4). The task in hand is thus to construct the simultaneous eigenfunctions of the two eigenvalue equations, i.e. the expressions for the $u(jm)$ in the r representation.[9]

The solution of the equation

$$L_z \psi(\theta, \varphi) = -i\hbar \frac{\partial}{\partial \varphi} \psi(\theta, \varphi) = \lambda \psi(\theta, \varphi)$$

is clearly $\lambda/\hbar = m = 0, \pm 1, \pm 2, \ldots$ and $\psi(\theta, \varphi) = a(\theta) \exp im\varphi$. m is restricted to integer values since ψ must be a single-valued function of φ. We now suppose that the function $\psi(\theta, \varphi)$ is an eigenfunction of L^2 with eigenvalue $\hbar^2 l(l+1)$ and of L_z with eigenvalue $\hbar m$, $|m| \leq l$ and write it as

(2.5.1) $$\psi_{lm}(\theta) \exp im\varphi.$$

The eigenfunction $\psi_{l,-l}(\theta)$ and in succession all the other eigenfunctions $\psi_{l,-l+1}(\theta), \ldots$ may be constructed by application of the differential operators (cf. (2.1.4))

(2.5.2) $$L_+ = L_x + iL_y = \hbar \exp i\varphi \left(\frac{\partial}{\partial \theta} + i \cot \theta \frac{\partial}{\partial \varphi} \right)$$

---

[9] Cf. Schiff (1949) p. 130.

and

(2.5.3) $\quad L_- = L_x - iL_y = \hbar \exp - i\varphi \left(-\dfrac{\partial}{\partial \theta} + i \cot \theta \dfrac{\partial}{\partial \varphi} \right)$

We compare the results of these applications with the expressions

$$L_+ u(l\ m) = \hbar[(l-m)(l+m+1)]^{\frac{1}{2}} u(l\ m+1)$$

and

$$L_- u(l\ -l) = 0$$

obtained by reference to the matrices of the angular momentum operators (2.3.16) and (2.3.17), so that the members of the set of eigenfunctions are correctly related to each other with respect to phase and normalization. The overall normalization will be specified later.

The above equations give immediately

$$\dfrac{\partial}{\partial \theta} \psi_{l\ -l}(\theta) - l \cot \theta\ \psi_{l\ -l}(\theta) = 0$$

from which we get

$$\log \psi_{l-l}(\theta) = l \log \sin \theta + C$$

i.e.

$$\psi_{l-l}(\theta) = a(\sin \theta)^l$$

where a is independent of θ and φ.

We have also

$$L_+ \psi_{lm}(\theta) \exp im\varphi = \hbar \left(\dfrac{\partial}{\partial \theta} - m \cot \theta \right) \psi_{lm}(\theta) \exp i(m+1)\varphi$$

$$= \hbar[(l-m)(l+m+1)]^{\frac{1}{2}} \psi_{l\ m+1}(\theta) \exp i(m+1)\varphi.$$

Hence

$$\psi_{l\ m+1}(\theta) = [(l-m)(l+m+1)]^{-\frac{1}{2}} \left(\dfrac{d}{d\theta} - m \cot \theta \right) \psi_{lm}(\theta)$$

We apply this relation $(l+m)$ times to the known $\psi_{l\ -l}(\theta)$ to obtain, after rearrangement,

$$\psi_{lm}(\theta) = (-1)^{l+m} a \left[\dfrac{(l-m)!}{(2l)!(l+m)!} \right]^{\frac{1}{2}} (\sin \theta)^m \left(\dfrac{d}{d \cos \theta} \right)^{l+m} (\sin \theta)^{2l}$$

Now we wish to normalize our eigenfunctions so that the integral of the probability over the sphere is unity. This implies for the eigenfunction $\psi_{l\ -l}(\theta, \varphi)$ that

$$\int_0^{2\pi} \int_0^{\pi} \psi^*_{l-l}(\theta, \varphi) \psi_{l\ -l}(\theta, \varphi) \sin \theta\ d\theta\ d\varphi = 1$$

i.e.

$$2\pi a^2 \int_0^\pi (\sin\theta)^{2l+1}\, d\theta = 1$$

Hence

$$a = \frac{1}{2^l l!}\left[\frac{(2l+1)!}{4\pi}\right]^{\frac{1}{2}}$$

We define $Y_{lm}(\theta,\varphi)$ as the normalized eigenfunction, and have therefore

(2.5.4) $$\int_0^{2\pi}\int_0^\pi Y^*_{lm}(\theta,\varphi) Y_{l'm'}(\theta,\varphi) \sin\theta\, d\theta\, d\varphi = \delta_{ll'}\delta_{mm'}$$

and

(2.5.5) $$Y_{lm}(\theta,\varphi) = \frac{(-1)^{l+m}}{2^l l!}\left[\frac{(2l+1)(l-m)!}{4\pi(l+m)!}\right]^{\frac{1}{2}}$$
$$\cdot (\sin\theta)^m \left[\frac{\partial}{\partial(\cos\theta)}\right]^{l+m} (\sin\theta)^{2l} \exp im\varphi$$

Application of Leibnitz' theorem to the expressions for $Y_{lm}(\theta,\varphi)$ and $Y_{l-m}(\theta,\varphi)$ and comparison of the resulting series shows that

(2.5.6) $$Y_{l-m}(\theta,\varphi) = (-1)^m Y^*_{lm}(\theta,\varphi)$$

It is not always convenient to define the Y_{lm} so that they have the symmetry relation (2.5.6). We shall see in Chapter 3 that it is an advantage from some points of view to take the function which is defined by

(2.5.7) $$\mathcal{Y}_{lm}(\theta,\varphi) = (i)^l Y_{lm}(\theta,\varphi)$$

and which has the symmetry property

(2.5.8) $$\mathcal{Y}_{l-m}(\theta,\varphi) = (-1)^{l+m}\mathcal{Y}^*_{lm}(\theta,\varphi)$$

However Y_{lm} is the most commonly used convention (cf. Condon and Shortley (1935)). But the $Y_{lm}(\theta,\varphi)$ of Bethe (1933) differ from our Y_{lm} by $(-1)^m$ and the $Y_{lm}(\theta,\varphi)$ of Schiff (1949) are equal to our Y_{lm} for negative m and differ by $(-1)^m$ for positive m.

The $Y_{lm}(\theta,\varphi)$ may be expressed in terms of the *associated Legendre functions*, whose properties will now be discussed.

We have assumed that the $Y_{lm}(\theta,\varphi)$ are solutions of the eigenvalue equation $\mathbf{L}^2\psi = \hbar^2\lambda\psi$ i.e.

$$-\hbar^2\left[\frac{1}{\sin\theta}\frac{\partial}{\partial\theta}\left(\sin\theta\frac{\partial}{\partial\theta}\right) + \frac{1}{\sin^2\theta}\frac{\partial^2}{\partial\varphi^2}\right]\psi = \hbar^2\lambda\psi$$

If we take $\psi = \psi_m(\theta)\exp im\varphi$ we have

$$-\hbar^2\left[\frac{1}{\sin\theta}\frac{d}{d\theta}\left(\sin\theta\frac{d}{d\theta}\right) - \frac{m^2}{\sin^2\theta}\right]\psi_m(\theta) = \hbar^2\lambda\psi_m(\theta)$$

The same equation appears when we separate the Laplace equation or the wave equation in spherical coordinates:

$$(\Delta + k^2)\psi \equiv \left[\frac{1}{r^2}\frac{\partial}{\partial r}\left(r^2\frac{\partial}{\partial r}\right) + \frac{1}{r^2\sin\theta}\cdot\frac{\partial}{\partial\theta}\left(\sin\theta\frac{\partial}{\partial\theta}\right)\right.$$
$$\left. + \frac{1}{r^2\sin^2\theta}\cdot\frac{\partial^2}{\partial\varphi^2} + k^2\right]\psi = 0$$

taking a solution $\psi = R(r)\Theta(\theta)\Phi(\varphi)$ and the separation constants m^2 and λ. The $Y_{lm}(\theta,\varphi)$ are thus the *spherical harmonics*.

THE ASSOCIATED LEGENDRE FUNCTIONS. If we write $\cos\theta = x$ and $\lambda = l(l+1)$ we obtain Legendre's differential equation

(2.5.9) $\quad (1-x^2)\dfrac{d^2y}{dx^2} - 2x\dfrac{dy}{dx} + \left[l(l+1) - \dfrac{m^2}{1-x^2}\right]y = 0$

It is known that for x real and between -1 and $+1$, this equation has one-valued continuous solutions for l and m integers. It is no restriction to assume that l is a non-negative integer. The solution which is finite at all points x, $-1 \leq x \leq +1$, which is the one we require, is only nonzero for $|m| \leq l$. The equation is satisfied by the associated Legendre functions of the first and second kinds,

(2.5.10) $\quad P_l^m(x) = (1-x^2)^{m/2}\dfrac{d^m P_l(x)}{dx^m}\ ;\quad Q_l^m(x) = (1-x^2)^{m/2}\dfrac{d^m Q_l(x)}{dx^m}$

where m is a positive integer and $P_l(x)$, $Q_l(x)$ are the Legendre functions. (We shall not be concerned with the second solution $Q_l^m(x)$ which is not finite at all points x, $-1 \leq x \leq +1$). This definition is that of *Ferrers*, and is used when the argument is real. The definition of *Hobson*, namely

(2.5.11) $\quad P_l^m(z) = (z^2-1)^{m/2}\dfrac{d^m P_l(z)}{dz^m}\ ;\quad Q_l^m(z) = (z^2-1)^{m/2}\dfrac{d^m Q_l(z)}{dz^m}$

is used when the argument is complex;[10] it is not employed in this book.

The *Legendre polynomials* $P_l(x)$ are defined by the generating function

(2.5.12) $\quad (1 - 2xh + h^2)^{-\frac{1}{2}} = P_0(x) + hP_1(x) + h^2 P_2(x) + \cdots$

whence

$$P_0(x) = 1,\quad P_1(x) = x,\quad P_2(x) = \tfrac{1}{2}(3x^2-1),$$
$$P_3(x) = \tfrac{1}{2}(5x^3-3x),\ \cdots$$

and generally

---

[10] Cf. Hobson (1931) p. 93, Whittaker and Watson (1946) pp. 323–5. Centre National d'Études des Télécommunications (1952).

2.5 · EIGENVECTORS OF THE OPERATORS

(2.5.13)
$$P_l(x) = \sum_{r=0}^{\nu} (-1)^r \frac{(2l-2r)! x^{l-2r}}{2^l r!(l-r)!(l-2r)!}$$
$$= {}_2F_1\left(-l, l+1, 1; \frac{1-x}{2}\right)$$

where $\nu = l/2$ or $(l-1)/2$, whichever is integer. In particular

$$P_l(1) = 1, \quad P_l(-x) = (-1)^l P_l(x)$$

The polynomials are given by the Rodrigues formula

(2.5.14)
$$P_l(x) = \frac{1}{2^l l!} \frac{d^l}{dx^l} (x^2 - 1)^l$$

and satisfy the differential equation

(2.5.15)
$$(1-x^2)\frac{d^2y}{dx^2} - 2x\frac{dy}{dx} + l(l+1)y = 0$$

The second solution of this equation is the Legendre function $Q_l(x)$. The Legendre polynomials are orthogonal:

(2.5.16)
$$\int_{-1}^{+1} P_k(x) P_l(x)\, dx = \frac{2\delta_{kl}}{2k+1}$$

a result which may be obtained by integrating by parts, making use of the Rodrigues formula.

We follow Hobson (1931) p. 99 and Bateman (1932) p. 361,[11] and generalize the definition of the associated Legendre function $P_l^m(x)$ to include negative values of m. Equations (2.5.10) and (2.5.14) are combined to give

(2.5.17)
$$P_l^m(x) = \frac{(1-x^2)^{m/2}}{2^l l!} \frac{d^{l+m}}{dx^{l+m}} (x^2 - 1)^l$$

We suppose this relation to define the $P_l^m(x)$ for $m = -1, -2, \ldots, -l$. Application of Leibnitz' theorem to the appropriate Rodrigues formulas shows that

(2.5.18)
$$P_l^{-m}(x) = (-1)^m \frac{(l-m)!}{(l+m)!} P_l^m(x)$$

The Rodrigues formula also shows that

(2.5.19)
$$P_l^m(-x) = (-1)^{l+m} P_l^m(x)$$

The following difference equations are satisfied by the $P_l^m(x)$

(2.5.20) $(l - m + 1)P_{l+1}^m(x) - (2l + 1)x P_l^m(x) + (l + m)P_{l-1}^m(x) = 0$

---
[11]See also Darwin (1928).

(2.5.21) $\quad xP_l^m(x) - (l - m + 1)(1 - x^2)^{\frac{1}{2}}P_l^{m-1}(x) - P_{l-1}^m(x) = 0$

(2.5.22) $\quad P_{l+1}^m(x) - xP_l^m(x) - (l + m)(1 - x^2)^{\frac{1}{2}}P_l^{m-1}(x) = 0$

(2.5.23) $\quad (l - m + 1)P_{l+1}^m(x) + (1 - x^2)^{\frac{1}{2}}P_l^{m+1}(x)$
$$- (l + m + 1)xP_l^m(x) = 0$$

(2.5.24) $\quad (1 - x^2)^{\frac{1}{2}}P_l^{m+1}(x) - 2mxP_l^m(x)$
$$+ (l + m)(l - m + 1)(1 - x^2)^{\frac{1}{2}}P_l^{m-1}(x) = 0$$

(2.5.25) $\quad (1 - x^2)\dfrac{d}{dx}P_l^m(x) = (l + 1)xP_l^m(x) - (l - m + 1)P_{l+1}^m(x)$
$$= (l + m)P_{l-1}^m(x) - lxP_l^m(x)$$

Important integral relations are

(2.5.26) $\quad \displaystyle\int_{-1}^{1} P_k^m(x)P_l^m(x)\, dx = \dfrac{2\delta_{kl}(l + m)!}{(2l + 1)(l - m)!}$

(2.5.27) $\quad \displaystyle\int_{-1}^{1} P_l^m(x)P_l^n(x)\, \dfrac{dx}{1 - x^2} = \dfrac{\delta_{mn}(l + m)!}{m(l - m)!}$

The first term of the asymptotic expansion[12] of $P_l^m(\cos\theta)$ for large l is given by

(2.5.28)
$$P_l^m(\cos\theta)$$
$$= (-1)^m \left(\dfrac{2}{\pi l \sin\theta}\right)^{\frac{1}{2}} \cos\left[\left(l + \dfrac{1}{2}\right)\theta - \dfrac{\pi}{2} + \dfrac{m\pi}{2}\right] + O(l^{-\frac{3}{2}})$$

where
$$\varepsilon \le \theta \le \pi - \varepsilon, \quad \varepsilon > 0, \quad l \gg m, \quad l \gg \dfrac{1}{\varepsilon}$$

RELATIONS BETWEEN THE EIGENFUNCTIONS Y_{lm} AND THE ASSOCIATED LEGENDRE FUNCTIONS. Comparison of the definition of the $Y_{lm}(\theta, \varphi)$ and the Rodrigues formula for the $P_l^m(\cos\theta)$ shows that the functions are related by

(2.5.29) $\quad Y_{lm}(\theta, \varphi) = (-1)^m \left[\dfrac{(2l + 1)(l - m)!}{4\pi(l + m)!}\right]^{\frac{1}{2}} P_l^m(\cos\theta) \exp im\varphi$

In particular we have

(2.5.30) $\quad Y_{l0}(\theta, \varphi) = \left(\dfrac{2l + 1}{4\pi}\right)^{\frac{1}{2}} P_l(\cos\theta)$

We shall find when we come to deal with tensor operators, that to avoid annoying factors it is convenient to use the notation of Racah

---

[12] Erdélyi (1953) §3.9.1.

(1942), namely to define

(2.5.31) $$C_q^{(k)} = \left(\frac{4\pi}{2k+1}\right)^{\frac{1}{2}} Y_{kq}(\theta, \varphi)$$

2.6. The Spin Eigenvectors

It was shown in (2.3) that representations of the angular momentum operators exist for half-odd-integer values of j and m. The basis vectors $u(j\,m)$ for such representations may not be expressed in terms of single-valued continuous functions on a sphere, as can the $u(j\,m)$ for integer j and m. We must therefore be content to consider them as quantities which have certain transformation properties under infinitesimal rotations, and which are normalized according to (2.3.3); the scalar product of the eigenvectors is no longer supposed to be associated with an integration over configuration space, as in (2.5.4).

There is, however a useful notation for these eigenvectors which is frequently employed, namely to write them as *column vectors*. Let us take an arbitrary linear combination v of a set of $2j+1$ eigenvectors $u(j\,m)$, which form the basis of a representation $\mathfrak{D}^{(j)}$.

$$v = \sum_m u(j\,m)(m|v)$$

If J is an angular momentum operator, we have

$$v' = Jv = \sum_{mm'} u(j\,m')(j\,m'|J|j\,m)(m|v)$$

The new coefficients $(m'|v')$ are thus given by

$$(m'|v') = \sum_m (j\,m'|J|j\,m)(m|v)$$

That is, the coefficients transform *contragrediently* to the eigenvectors $u(j\,m)$, and the set of coefficients $(m|v)$ belonging to a vector v may be represented as a column vector from the point of view of matrix multiplication. In this scheme an eigenvector $u(j\,m)$ will appear as

$$\begin{bmatrix} \delta_{j,m} \\ \delta_{j-1,m} \\ \vdots \\ \delta_{-j,m} \end{bmatrix}$$

(We suppose, as always, that the m values labeling rows and columns in a matrix decrease from left to right and from top to bottom.) In particular the eigenvectors $u(\tfrac{1}{2}\,\tfrac{1}{2})$ and $u(\tfrac{1}{2}\,-\tfrac{1}{2})$ may be written as

(2.6.1) $$u(\tfrac{1}{2}\,\tfrac{1}{2}) \sim \begin{pmatrix} 1 \\ 0 \end{pmatrix}; \quad u(\tfrac{1}{2}\,-\tfrac{1}{2}) \sim \begin{pmatrix} 0 \\ 1 \end{pmatrix}$$

and a general spin vector as

$$\begin{pmatrix} \psi_+ \\ \psi_- \end{pmatrix}$$

If the spin is a function of position, this column vector will appear as

$$\begin{pmatrix} \psi_+(\mathbf{r}) \\ \psi_-(\mathbf{r}) \end{pmatrix}$$

where

$$\int \{|\psi_+(\mathbf{r})|^2 + |\psi_-(\mathbf{r})|^2\}\, d\mathbf{r} = 1$$

Thus we may consider a spin $\frac{1}{2}$ particle to be described by a *pair* of functions on the configuration space.

DIFFERENTIAL OPERATORS IN SPIN SPACE. The eigenvectors $u(\frac{1}{2}\,\frac{1}{2})$, $u(\frac{1}{2}\,-\frac{1}{2})$ define a linear unitary space of two dimensions, the so-called spin space; and the transformations corresponding to the matrices (2.3.19) may be considered to be equivalent to certain differential operators in this space. We shall henceforth write for conciseness

(2.6.2) $\qquad u(\tfrac{1}{2}\,\tfrac{1}{2}) \equiv \chi_+; \qquad u(\tfrac{1}{2}\,-\tfrac{1}{2}) \equiv \chi_-$

and the differential operators

(2.6.3) $\qquad \dfrac{\partial}{\partial \chi_+} \equiv \partial_+, \qquad \dfrac{\partial}{\partial \chi_-} \equiv \partial_-$

It is easy to see that in the $\mathfrak{D}^{(\frac{1}{2})}$ representation we may equate the angular momentum operators with linear differential operators in the following way; i.e. the results of operating with the quantities (2.6.4) on the χ's given by (2.6.2) correspond to the results of operating with the matrices (2.3.19):

$$J_x \sim \frac{\hbar}{2}(\chi_- \partial_+ + \chi_+ \partial_-)$$

$$J_y \sim \frac{i\hbar}{2}(\chi_- \partial_+ - \chi_+ \partial_-)$$

(2.6.4)

$$J_z \sim \frac{\hbar}{2}(\chi_+ \partial_+ - \chi_- \partial_-)$$

$$J_+ \sim \hbar \chi_+ \partial_-$$

$$J_- \sim \hbar \chi_- \partial_+$$

The square of the total angular momentum appears as

$$\mathbf{J}^2 = J_z(J_z - \hbar) + J_+ J_-$$

(2.6.5) $\quad = \dfrac{\hbar^2}{4}(\chi_+\chi_+\partial_+\partial_+ + \chi_-\chi_-\partial_-\partial_- + 2\chi_+\chi_-\partial_+\partial_- + 3\chi_+\partial_+ + 3\chi_-\partial_-)$

$\quad = \hbar^2 k(k+1) \quad \text{where} \quad k = \tfrac{1}{2}(\chi_+\partial_+ + \chi_-\partial_-)$

These spinor differential operator expressions lead us to a new and useful way of representing the angular momentum eigenvectors. Let us consider an arbitrary monomial in the χ_+, χ_-, say[13]

$$\chi_+^x \chi_-^y$$

Then it is clearly a simultaneous eigenvector of J_z and $\mathbf{J}^2$ when they are expressed in the form (2.6.4) and (2.6.5). Moreover, the eigenvalues are

$$\frac{\hbar}{2}(x-y) \quad \text{and} \quad \hbar^2\left(\frac{x+y}{2}\right)\left(\frac{x+y}{2}+1\right)$$

respectively. The result of operation with J_+ or J_- is to change the values of x and y but to leave the degree $x+y$ unchanged. These facts imply that the set of $2j+1$ monomials $\chi_+^{j+m}\chi_-^{j-m}$ where $m = -j, -j+1, \ldots, j-1, j$ form a basis for the $\mathfrak{D}^{(j)}$ representation of the angular momentum operators. If we normalize these monomials by writing

(2.6.6) $\quad u(j\,m) \equiv \dfrac{\chi_+^{j+m}\chi_-^{j-m}}{+[(j+m)!(j-m)!]^{\frac{1}{2}}}$

we find that their behavior under application of the angular momentum operators in the form (2.6.4) follows exactly that of the $u(j\,m)$ in (2.3.15), (2.3.16), and (2.3.17). This representation arises from the correspondence between the rotation group $SO(3)$ and the group of unitary unimodular (determinant $+1$) 2×2 matrices $SU(2)$. The reader is referred to works on group theory for further details. See for example Eckart (1930), Weyl (1931), Van der Waerden (1931), Bauer (1933).

2.7. Angular Momentum Eigenfunctions in the Case of Large l

We shall examine, by means of the WKB method, the behavior of the angular momentum eigenfunctions $Y_{lm}(\theta, \varphi)$ when l is large. The most significant result will be that the probability density $|\psi(\theta, \varphi)|^2$, apart from rapid oscillations, approaches that of a classical particle moving in a circular orbit. The substitutions

$$a = \sin\theta = \left[1 - \frac{m^2}{l(l+1)}\right]^{\frac{1}{2}}, \quad w = [1-x^2]^{\frac{1}{4}}y,$$

$$\varepsilon = [l(l+1)]^{-\frac{1}{2}}, \quad \cos\theta = \frac{m}{[l(l+1)]^{\frac{1}{2}}}$$

---

[13]Such an expression may be considered as derived from a symmetric state vector describing $(x+y)$ spin $\tfrac{1}{2}$ particles; such a system has no physical significance in itself since spin $\tfrac{1}{2}$ particles obey Fermi-Dirac statistics. Cf. Pauli (1941).

are made in the Legendre equation (2.5.9), and furnish the equation

$$(2.7.1) \qquad \varepsilon^2 \frac{d^2 w}{dx^2} + \left(\frac{a^2 - x^2 + \varepsilon^2}{(1 - x^2)^2}\right) w = 0$$

APPLICATION OF THE WKB METHOD. Putting $w = \exp(iS(x)/\varepsilon)$ we obtain

$$(2.7.2) \qquad i\varepsilon \frac{d^2 S}{dx^2} - \left(\frac{dS}{dx}\right)^2 + \frac{a^2 - x^2 + \varepsilon^2}{(1 - x^2)^2} = 0$$

The reasoning of the WKB method (cf. Schiff (1949) p. 178) shows that if

$$(2.7.3) \qquad \left| \frac{\varepsilon \dfrac{d}{dx}\left[\dfrac{a^2 - x^2 + \varepsilon^2}{(1 - x^2)^2}\right]^{\frac{1}{2}}}{2\left(\dfrac{a^2 - x^2 + \varepsilon^2}{(1 - x^2)^2}\right)} \right| \ll 1, \quad \text{then}$$

either

$$(2.7.4) \qquad w(x) \cong A \frac{(1 - x^2)^{\frac{1}{2}}}{(a^2 - x^2)^{\frac{1}{4}}} \exp \pm \frac{i}{\varepsilon} \int^x \frac{(a^2 - x^2)^{\frac{1}{2}}}{1 - x^2} dx$$

or

$$w(x) \cong B \frac{(1 - x^2)^{\frac{1}{2}}}{(x^2 - a^2)^{\frac{1}{4}}} \exp \pm \frac{1}{\varepsilon} \int^x \frac{(x^2 - a^2)^{\frac{1}{2}}}{1 - x^2} dx$$

according to whether $a^2 + \varepsilon^2 > x^2$ or $a^2 + \varepsilon^2 < x^2$. I.e. in the oscillatory region ($a^2 + \varepsilon^2 > x^2$) we have

$$(2.7.5) \qquad Y_{lm}(\theta, \varphi) \cong \frac{A}{(a^2 - x^2)^{\frac{1}{4}}} \exp \pm \frac{i}{\varepsilon} \int^x \frac{(a^2 - x^2)^{\frac{1}{2}}}{1 - x^2} dx \cdot \exp im\varphi$$

It is easy to see that when $|m| \ll l$, i.e. when $a \to 1$, the expression obtained for Y_{lm} is compatible with the first term of the asymptotic expansion (2.5.28) for $P_l^m(\cos \theta)$.

PROBABILITY DENSITY.* The probability density $P(\cos \theta)$ oscillates rapidly with $\cos \theta$ (note the number of zeros of $P_l^m(\cos \theta)$ is $l - |m|$), but if we consider the average of these oscillations, we have

$$(2.7.6) \qquad \overline{P(\cos \theta)} = \int |Y_{lm}(\theta, \varphi)|^2 d\varphi \cong \frac{A}{(\sin^2 \Theta - \cos^2 \theta)^{\frac{1}{2}}},$$

except in the neighborhood of $\cos \theta = \sin \Theta$, where the expression on the right becomes infinite, while $\overline{P(\cos \theta)}$ does not. This expression on the right is the classical density distribution in $\cos \theta$ of a point particle moving in a circular orbit about the origin, the axis of the orbit making an angle Θ with the z-axis. (The density is inversely proportional to the quantity $d(\cos \theta)/dt$.) The classical density distri-

---

*See Brussaard and Tolhoek (1957).

bution is zero for $\theta < \pi/2 - \Theta$, while $\overline{P(\cos\theta)}$ has a finite value in this region, decreasing roughly exponentially to zero as θ decreases. Clearly the larger l, the closer $\overline{P(\cos\theta)}$ will approach the classical distribution.

UNCERTAINTY IN DIRECTION OF ANGULAR MOMENTUM VECTOR. The quantum mechanical probability density, not being time dependent, gives us no information about the motion of the particle in its orbit. Moreover we have no information about the coordinate of the axis of rotation; it is as if the orbit "precessed" in an unobservable way about the z-axis.

REPRESENTATION OF AN ANGULAR MOMENTUM WHOSE DIRECTION IS WELL DEFINED. We note, however, that if $m = l$ the direction of the angular momentum vector is relatively well defined (cf. (2.4)). In this case we have $\int |Y_{ll}(\theta, \varphi)|^2 \, d\varphi \cong A(\sin\theta)^{2l}$ and for large l the probability distribution is that of a particle moving in a well-defined orbit whose axis is the z-axis.

It is possible to represent such an orbit, or such a system with a well-defined angular momentum vector with any orientation simply by carrying out a unitary transformation which corresponds to a rotation of coordinates from one S' whose z-axis coincides with the angular momentum axis to the actual coordinate system S. In the system S' the orbit is represented by $Y_{ll}(\theta, \varphi)$. The transformation to eigenvectors defined in the system S is

$$(2.7.7) \qquad \sum_m Y_{lm}(\theta, \varphi) \mathfrak{D}^{(l)}_{ml}(\alpha\,\beta\,\gamma)$$

where $\alpha\,\beta\,\gamma$ are the Euler angles associated with the rotation of axes and the coefficients $\mathfrak{D}$ are the matrix elements of finite rotations (cf. Chapter 4). In the sufficiently typical case when the axis points in a direction in the x, z plane whose angle with the z-axis is β, we have (see Eq. (4.1.27)) for the above series

$$(2.7.8) \qquad \sum_m Y_{lm}(\theta, \varphi) \left[\frac{(2l)!}{(l+m)!(l-m)!}\right]^{\frac{1}{2}} \left(\cos\frac{\beta}{2}\right)^{l+m} \left(\sin\frac{\beta}{2}\right)^{l-m}.$$

The indeterminacy in the value of m in such a state vector is, of course, associated with the impossibility of measuring φ and $L_z = -i\hbar(\partial/\partial\varphi)$ simultaneously.

2.8. Time Reversal and the Angular Momentum Operators

We shall use in the following chapters a number of properties of the so-called *time-reversed* angular momentum operators. The properties of the operators of *orbital* angular momentum under time reversal, i.e. the replacement of t by $-t$, are easily found. The definition of **L** in terms of **r** and **p** shows that L_x, L_y, and L_z must be replaced by $-L_x$,

$-L_y$, and $-L_z$ respectively. The orbital angular momentum eigenfunctions, being associated with solutions of a Schrödinger equation, may be simply replaced by their complex conjugates.

However the properties of the spin operators and eigenvectors under time reversal are not so evident; the reader is referred to the paper of Wigner (1932), who shows that the operation of time reversal when spin is involved must correspond to a unitary operator U accompanied by a complex conjugation K_0. I.e.

(2.8.1) $$K = UK_0$$

We shall examine the set of eigenvectors $\tilde{u}(j\,m)$ associated with the time reversed angular momentum operators $KJ_xK = -J_x$, $KJ_yK = -J_y$, $KJ_zK = -J_z$ (denoted by $\tilde{J}_x$, $\tilde{J}_y$, $\tilde{J}_z$), which are analogous to the eigenvectors $u(j\,m)$ associated with J_x, J_y, and J_z. Since $\tilde{\mathbf{J}}^2 = \mathbf{J}^2$ and $\tilde{J}_z = -J_z$ we have

$$\tilde{u}(j\,m) = \alpha(j\,m)u(j\,-m)$$

The matrices of $\tilde{J}_x$ and $\tilde{J}_y$ are obtained in the same way as those of J_x and J_y; and we find a consistent scheme when we take

(2.8.2) $$\tilde{u}(j\,m) = (-1)^{j+m}u(j\,-m)$$

The relation is arbitrary within a phase independent of m; the choice above corresponds, in the case of integer j, to the phase of the function $\mathfrak{Y}_{lm}$ (2.5.8).

CHAPTER 3

The Coupling of Angular Momentum Vectors

3.1. The Addition of Angular Momenta

DISCUSSION OF A CLASSICAL MODEL. The total angular momentum of a classical mechanical system composed of two parts, each having an angular momentum whose magnitude and direction are well defined, is easily obtained. It is given by the vector **L** which is the resultant of the addition of the two vectors $\mathbf{L}_1$ and $\mathbf{L}_2$. However we have seen in Chapter 2 that in quantum mechanics, even in the limit of large angular momenta, we do not specify the angular momentum of a system in such a way that we may speak of the direction of the angular momentum vector; we know only the magnitude of the vector and its projection on a given axis. It is therefore instructive to consider the derivation of the angular momentum of a classical system which corresponds to this situation; we take two vectors $\mathbf{L}_1$ and $\mathbf{L}_2$ whose lengths l_1, l_2 and whose projections m_1, m_2 on the z-axis are fixed, but whose orientations are otherwise undefined; in fact we suppose the angles φ_1, φ_2 to take equally probably any values between 0 and 2π. It follows that the resultant angular momentum vector **L** has a probability distribution over a range of lengths and orientations. An elementary application of the principle of vector addition (see Fig. 3.1) shows that (i) the projection m on the z-axis is fixed, being the sum of the projections of the vectors $\mathbf{L}_1$ and $\mathbf{L}_2 : m = m_1 + m_2$; (ii) the length l of the vector and correspondingly the angle φ with the z-axis, must fall within a range of values. The bounds on this range depend on the values of m_1 and m_2, but must always be consistent with the requirement $|l_1 - l_2| \leq l \leq l_1 + l_2$; (iii) the angle φ may take equally probably any value between 0 and 2π.

We may compute the probability density $P(l)$ for l; (i.e. the probability that the length of **L** lies between l and $l + dl$ is $P(l)\,dl$). If we suppose $\mathbf{L}_2$ to rotate at a constant rate about the z-axis with respect to $\mathbf{L}_1$, we have that $P(l)$ is inversely proportional to dl/dt. We have in fact

$$l^2 = m^2 + l_1^2 \sin^2 \theta_1 + l_2^2 \sin^2 \theta_2 + 2l_1 l_2 \sin \theta_1 \sin \theta_2 \cos(\varphi_1 - \varphi_2)$$

where θ_1, θ_2 are the angles made by $\mathbf{L}_1$, $\mathbf{L}_2$ with the z-axis. Hence

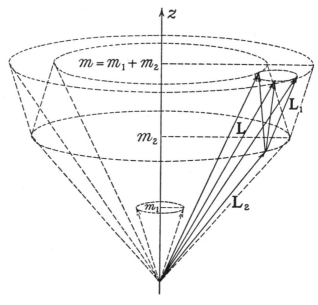

Fig. 3.1

$$P(l) \sim \left(\frac{dl}{dt}\right)^{-1} \sim 2l[2l^2(m^2 + l_1^2 \sin^2 \theta_1 + l_2^2 \sin^2 \theta_2) + 2l_1^2 l_2^2 \sin^2 \theta_1 \sin^2 \theta_2$$
$$- l^4 - m^4 - l_1^4 \sin^4 \theta_1 - l_2^4 \sin^4 \theta_2$$
(3.1.1)
$$- 2m_1^2 l_1^2 \sin^2 \theta_1 - 2m_2^2 l_2^2 \sin^2 \theta_2]^{-\frac{1}{2}}$$

The probability density becomes infinite at the bounding values of l and varies smoothly between them.

When we go over to quantum mechanics, we replace our continuously varying probability density $P(l)$ by the squares of the coefficients in the series of terms obtained by analyzing a state specified by the quantum numbers j_1, m_1; j_2, m_2 into states specified by the quantum numbers j, m of total angular momentum. Although we shall go about computing the coefficients in a quite different way from that in which we obtained $P(l)$, we shall find their behavior (allowing for the fact that the angular momenta take discrete values) similar to that of $P(l)$, especially when large values of angular momenta are considered.

We have supposed up to now that there is no interaction between the two parts of the system; if we introduce an interaction it is well known that, although the total angular momentum **L** will be unaltered, the vectors **L₁** and **L₂** will precess about the axis of **L**. Now this is a quite well-defined statement when the orientation of the vectors is specified; however in our model it corresponds only to the values of m_1 and m_2 varying over a certain range, the sum $m_1 + m_2$ remaining constant. We shall see that this corresponds in quantum mechanics to the case where an interaction connects states of different m_1, m_2, these quantities being no longer good quantum numbers.

THE QUANTUM MECHANICAL PROBLEM. We base our treatment on the operator equation

(3.1.2) $$\mathbf{J} = \mathbf{J}_1 + \mathbf{J}_2$$

The operators $\mathbf{J}_1$ and $\mathbf{J}_2$ commute, for they refer to independent systems; this implies that the components of $\mathbf{J}$ obey the commutation relations (2.1.2).

We shall study the unitary transformation which expresses the simultaneous eigenvectors $v(\gamma j_1 m_1 j_2 m_2)$ of the complete set of commuting operators Γ, $\mathbf{J}_1^2$, J_{1z}, $\mathbf{J}_2^2$, J_{2z} in terms of the simultaneous eigenvectors of the similar set Γ, $\mathbf{J}_1^2$, $\mathbf{J}_2^2$, $\mathbf{J}^2$, J_z (Γ represents the other operators in the complete set which, due to their invariance under rotation, do not enter into the discussion). It will be noticed that $\mathbf{J}_1^2$ and $\mathbf{J}_2^2$ appear in each set; they commute with $\mathbf{J}^2$ and J_z as well as with J_{1z} and J_{2z}, as is easily shown.

We have from (3.1.2) that

$$J_z = J_{1z} + J_{2z}$$

This implies immediately that the magnetic quantum numbers satisfy

(3.1.3) $$m = m_1 + m_2$$

a result identical with that obtained in the classical case.

We now consider the values of j which arise from particular values of j_1, j_2. We first express $\mathbf{J}^2$ in terms of the original operators:

(3.1.4) $$\begin{aligned}\mathbf{J}^2 &= (J_{1x} + J_{2x})^2 + (J_{1y} + J_{2y})^2 + (J_{1z} + J_{2z})^2 \\ &= \mathbf{J}_1^2 + \mathbf{J}_2^2 + 2(\mathbf{J}_1 \cdot \mathbf{J}_2) \\ &= \mathbf{J}_1^2 + \mathbf{J}_2^2 + J_{1+}J_{2-} + J_{1-}J_{2+} + 2J_{1z}J_{2z}\end{aligned}$$

It will be noticed that $\mathbf{J}^2$ connects states with different m_1 and m_2. Let us apply $\mathbf{J}^2$ to the state $v(\gamma j_1 j_1 j_2 j_2)$, i.e. the state with the maximum values of m_1 and m_2. We see from (3.1.4) that this state is an eigenstate of $\mathbf{J}^2$, with eigenvalue

$$\begin{aligned}\hbar^2 j(j+1) &= \hbar^2\{j_1(j_1+1) + j_2(j_2+1) + 2j_1j_2\} \\ &= \hbar^2(j_1+j_2)(j_1+j_2+1)\end{aligned}$$

I.e. $j = j_1 + j_2$, and we have

$$v(\gamma j_1 j_1 j_2 j_2) = e^{i\delta} w(\gamma j_1 j_2 j_1+j_2 \; j_1+j_2) \qquad (\delta \text{ real})$$

where $w(\gamma j_1 j_2 j m)$ is an eigenvector of Γ, $\mathbf{J}_1^2$, $\mathbf{J}_2^2$, $\mathbf{J}^2$, J_z; (the relative phase is as yet unspecified). Now we consider the states $v(\gamma\ j_1\ j_1-1\ j_2\ j_2)$ and $v(\gamma\ j_1\ j_1\ j_2\ j_2-1)$. It is clear that these states are connected by $\mathbf{J}^2$. Now a certain linear combination of these two states corresponds to the state $w(\gamma\ j_1\ j_2\ j_1+j_2\ j_1+j_2-1)$. For (2.3.17) implies

$$e^{i\delta}J_-w(\gamma\ j_1\ j_2\ j_1+j_2\ j_1+j_2) = \hbar(2j_1+2j_2)^{\frac{1}{2}}w(\gamma\ j_1\ j_2\ j_1+j_2\ j_1+j_2-1)$$
$$= (J_{1-} + J_{2-})v(\gamma\ j_1\ j_1\ j_2\ j_2)$$
$$= \hbar\{(2j_1)^{\frac{1}{2}}v(\gamma\ j_1\ j_1-1\ j_2\ j_2) + (2j_2)^{\frac{1}{2}}v(\gamma\ j_1\ j_1\ j_2\ j_2-1)\}$$

One can construct a state orthogonal to this one, namely

$$(2j_2)^{\frac{1}{2}}v(\gamma\ j_1\ j_1-1\ j_2\ j_2) - (2j_1)^{\frac{1}{2}}v(\gamma\ j_1\ j_1\ j_2\ j_2-1)$$

Application of (3.1.4) shows that this state is an eigenstate of $\mathbf{J}^2$, the eigenvalue being $\hbar^2(j_1+j_2)(j_1+j_2-1)$, i.e.

$$j = j_1 + j_2 - 1$$

When $m = j_1 + j_2 - 2$ the states involved are $v(\gamma\ j_1\ j_1-2,\ j_2\ j_2)$, $v(\gamma\ j_1\ j_1-1\ j_2\ j_2-1)$, $v(\gamma\ j_1\ j_1\ j_2\ j_2-2)$. We can, in the same way as before, construct out of these 3 states two states with j values which have already been found, namely $w(\gamma\ j_1\ j_2,\ j_1+j_2,\ j_1+j_2-2)$, $w(\gamma\ j_1\ j_2,\ j_1+j_2-1,\ j_1+j_2-2)$. There remains one other state orthogonal to these two, which we could show by calculation to be an eigenstate of $\mathbf{J}^2$, with $j = j_1 + j_2 - 2$. However we can see that it could in any case not have $j = j_1 + j_2$ or $j_1 + j_2 - 1$, since this would imply the existence of *two* states with $j = j_1 + j_2$, $m = j_1 + j_2$ or two with $j = j_1 + j_2 - 1$, $m = j_1 + j_2 - 1$ which could be obtained by application of the operator J_+. Now we may go on in this way, reducing $m = m_1 + m_2$ by one at each step, and obtaining at each step one more new value of j in the sequence $j_1 + j_2, j_1 + j_2 - 1, j_1 + j_2 - 2, \ldots$.

This process continues until either $m_1 = -j_1$ or $m_2 = -j_2$. The sequence of possible values of j is thus

(3.1.5) $\quad j_1 + j_2, j_1 + j_2 - 1, j_1 + j_2 - 2, \cdots, |j_1 - j_2| + 1, |j_1 - j_2|$.

and to each value of j corresponds $2j + 1$ states with $m = j, j - 1, \ldots, -j$. The number of states in the two representations must be the same; in fact

$$\sum_{j=|j_1-j_2|}^{j_1+j_2} (2j+1) = (2j_1+1)(2j_2+1).$$

The result (3.1.5) corresponds to the weaker limits on l in the classical model discussed, and is identical with the addition rule for angular momenta found empirically by spectroscopists before the discovery of wave or matrix mechanics.

Now if two angular momenta commute, they must refer to different particles or to different properties of the same particle, e.g. the orbital and spin angular momenta of an electron. It follows that a state vector of the type $v(\gamma\, j_1\, m_1\, j_2\, m_2)$ may be split up into a sum of products of factors relating to the separate parts of the system:

(3.1.6) $\qquad v(\gamma\, j_1\, m_1\, j_2\, m_2) = \sum_{\alpha_1 \alpha_2} v_1(\alpha_1\, j_1\, m_1) v_2(\alpha_2\, j_2\, m_2)$

Thus from the point of view of the representations of the rotation group, the $v(\gamma\, j_1\, m_1\, j_2\, m_2)$ are linear combinations of the basis elements of the product representation $\mathfrak{D}^{(j_1)} \otimes \mathfrak{D}^{(j_2)}$. This representation, of dimension $(2j_1 + 1)(2j_2 + 1)$, is reducible, i.e. the representation space splits up into a number of invariant irreducible subspaces, each corresponding to one of the allowed values of j. The determination of the allowed values of j may be carried out by group theoretical methods (cf. Weyl (1931) p. 123, Eckart (1930) etc.).

3.2. Commutation Relations between Components of J_1, J_2 and J

The following commutation relations involving the components of J may be confirmed by substitution for these components according to (3.1.2), remembering that the components of J_1 commute with those of J_2.

The components of J satisfy the commutation relations (2.1.2):

(3.2.1) $\qquad [J_x, J_y] = i\hbar J_z;\quad [J_z, J_x] = i\hbar J_y;\quad [J_y, J_z] = i\hbar J_x.$

The following relations are also valid for the components J_2 and J.

(3.2.2)
$$[J_x, J_{1x}] = 0 \quad [J_x, J_{1y}] = i\hbar J_{1z} \quad [J_x, J_{1z}] = -i\hbar J_{1y}.$$
$$[J_y, J_{1y}] = 0 \quad [J_y, J_{1z}] = i\hbar J_{1x} \quad [J_y, J_{1z}] = -i\hbar J_{1z}$$
$$[J_z, J_{1z}] = 0 \quad [J_z, J_{1x}] = i\hbar J_{1y} \quad [J_z, J_{1y}] = -i\hbar J_{1x}$$

It is convenient to rewrite these using the non-Hermitian operators

(3.2.3)
$$J_+ = J_x + iJ_y, \quad J_{1+} = J_{1x} + iJ_{1y}$$
$$J_- = J_x - iJ_y, \quad J_{1-} = J_{1x} - iJ_{1y}, \quad \text{etc.}$$

We have then

(3.2.4)
$$[J_\pm, J_{1\pm}] = 0, \qquad [J_\pm, J_{1z}] = \mp\hbar J_{1\pm}$$
$$[J_\pm, J_{1\mp}] = \pm 2\hbar J_{1z}, \quad [J_z, J_{1\pm}] = \pm\hbar J_{1\pm}$$

3.3. Selection Rules for the Matrix Elements of J_1 and J_2

The commutation relations between the components of J_1 and J imply certain selection rules on the matrix elements of J_1 in the

$(\gamma\, j_1\, j_2\, j\, m)$ scheme. Similar rules of course apply to the matrix elements of J_2.

(3.3.1) *The matrix elements of J_{1z} are diagonal in m.*

We have from $[J_z, J_{1z}] = 0$ that

$$m'(j'\, m'|J_{1z}|j\, m) - (j'\, m'|J_{1z}|j\, m)m = 0$$

I.e. the matrix element is zero unless $m' - m = 0$.

(3.3.2) *The operator J_{1+} increases m by one*

$$J_{1-} \text{ decreases } m \text{ by one.}$$

The relation $[J_z, J_{1\pm}] = \pm\hbar J_{1\pm}$ gives

$$m'(j'\, m'|J_{1\pm}|j\, m) - (j'\, m'|J_{1\pm}|j\, m)m = \pm(j'\, m'|J_{1\pm}|j\, m).$$

Hence $m' - m = \pm 1$ is a necessary condition for $(j'\, m'|J_{1\pm}|j\, m)$ to be nonzero.

(3.3.3) *The matrix elements of the components of J_1 are zero if $|j' - j| > 1$.*

Suppose on the contrary that $j' = j + 1 + \lambda$, $\lambda > 0$. Then $(j'\, m'|J_{1+}|j\, m) = 0$ for $m' > j + 1$. But we have $[J_+, J_{1+}] = 0$. Therefore

$$(j'\, m'|J_+|j'\, m'-1)\cdot(j'\, m'-1|J_{1+}|j\, m)$$
$$- (j'\, m'|J_{1+}|j\, m+1)(j\, m+1|J_+|j\, m) = 0$$

and $(j'\, m'|J_{1+}|j\, m)$ is zero for all m, m'. It is easy to see from the relations $[J_-, J_{1+}] = -2\hbar J_{1z}$ and $[J_-, J_{1z}] = \hbar J_{1-}$ that all matrix elements of J_1 between $j' = j + 1 + \lambda$ and j must be zero. A similar proof applies for $j' < j - 1$.

3.4. The Choice of the Phases of the States $w(\gamma\, j_1\, j_2\, j\, m)$

We have seen how it is in principle possible to construct eigenvectors of $\mathbf{J}^2$ and J_z from linear combinations of eigenvectors of $\mathbf{J}_1^2$, J_{1z}, $\mathbf{J}_2^2$, J_{2z}. The phases of these new eigenvectors with respect to the original ones or with respect to each other have not as yet been specified.

The first choice of phase is that which is implicit in identifying the eigenvector of highest j and m with the eigenvector in the original scheme to which, as we have seen in (3.1), it corresponds uniquely:

(3.4.1) $\qquad w(\gamma\, j_1\, j_2\, j_1+j_2\, j_1+j_2) \equiv v(\gamma\, j_1\, j_1\, j_2\, j_2)$

We now have to relate the phases of the $w(\gamma\, j_1\, j_2\, j\, m)$ with different j. Note first that, since J_{1z} clearly does not commute with $\mathbf{J}^2$ it must connect states of different j. Following from the selection rule (3.3.3), the j values may only differ by one. In analogy with (2.3), we may make an arbitrary choice of phase for the nondiagonal matrix elements of J_{1z}.

Before this is done, we must show that all matrix elements of J_{1z} between states of given j and j' have the same phase. We have from (3.2.4) and (3.3.2) that

$$(j\ m+1|J_+|j\ m)(j\ m|J_{1+}|j'\ m-1)$$
$$- (j\ m+1|J_{1+}|j'\ m)(j'\ m|J_+|j'\ m-1) = 0$$

and the matrix elements of J_+ are by convention (2.3.16), real and positive. Hence all matrix elements of J_{1+} between states of given j and j' have the same phase. Equation (3.2.4) gives us also

$$(j\ m+1|J_+|j\ m)(j\ m|J_{1z}|j+1\ m)$$
$$- (j\ m+1|J_{1z}|j+1\ m+1)(j+1\ m+1|J_+|j+1\ m)$$
$$= -\hbar(j\ m+1|J_{1+}|j+1\ m)$$

I.e.

$$(j\ m+1|J_{1z}|j+1\ m+1) = [(j+1\ m+1|J_+|j+1\ m)]^{-1}$$
$$\times \{(j\ m|J_{1z}|j+1\ m)(j\ m+1|J_+|j\ m) + \hbar(j\ m+1|J_{1+}|j+1\ m)\}$$

If we take $m = -j-1$ in the above equation we shall find that $(j, -j|J_{1z}|j+1, -j)$ has the phase of the matrix elements of J_{1+} between $j; j+1$. Suppose this phase is real and positive. Then all matrix elements of J_{1z} between $j; j+1$ are real and non-negative.

Thus we are at liberty to prescribe the convention:

(3.4.2) *All matrix elements of J_{1z} which are nondiagonal in j are real and non-negative.*

Since $J_z = J_{1z} + J_{2z}$ has only zero matrix elements between states of different j, it follows that the corresponding matrix elements of J_{2z} are real and nonpositive. These conventions are identical with those of Condon and Shortley (1935).

3.5. The Vector-Coupling Coefficients

DEFINITION. The eigenvectors $w(\gamma\ j_1\ j_2\ j\ m)$ are given in terms of the $v(\gamma\ j_1\ m_1\ j_2\ m_2)$ by the unitary transformation (cf. Dirac (1935) §17)

(3.5.1) $\quad w(\gamma\ j_1\ j_2\ j\ m) = \sum_{m_1 m_2} v(\gamma\ j_1\ m_1\ j_2\ m_2)(j_1\ m_1\ j_2\ m_2|j_1\ j_2\ j\ m)$

It is clear that the additional quantum numbers γ need not enter the coefficients. The inverse transformation is

$$(3.5.2) \qquad v(\gamma\, j_1\, m_1\, j_2\, m_2) = \sum_{jm} w(\gamma\, j_1\, j_2\, j\, m)(j_1\, j_2\, j\, m | j_1\, m_1\, j_2\, m_2)$$

where the coefficients $(j_1\, j_2\, j\, m | j_1\, m_1\, j_2\, m_2)$ are the complex conjugates of the corresponding $(j_1\, m_1\, j_2\, m_2 | j_1\, j_2\, j\, m)$. We shall see later that the coefficients are, with the choice of phases already made, in fact real. They are called *vector-coupling*, *Wigner*, or *Clebsch-Gordan* coefficients; they form unitary matrices of dimension $(2j_1+1)(2j_2+1)$ with rows and columns being labelled by the pairs m_1, m_2 and j, m respectively.

The vector addition rules (3.1.3) and (3.1.5) imply that the coefficients are zero unless the conditions on j and m are satisfied.

UNITARY PROPERTIES. The unitary properties of the coefficient matrix for given j_1, j_2 are expressed by

$$(3.5.3) \qquad \sum_{jm} (j_1\, m_1'\, j_2\, m_2' | j_1\, j_2\, j\, m)(j_1\, j_2\, j\, m | j_1\, m_1\, j_2\, m_2) = \delta_{m_1' m_1} \delta_{m_2' m_2}$$

$$(3.5.4) \qquad \sum_{m_1 m_2} (j_1\, j_2\, j'\, m' | j_1\, m_1\, j_2\, m_2)(j_1\, m_1\, j_2\, m_2 | j_1\, j_2\, j\, m)$$
$$= \delta_{j' j} \delta_{m' m} \delta(j_1\, j_2\, j)$$

where $\delta(j_1\, j_2\, j) = 1$ if j satisfies (3.1.5) and is zero otherwise.

Since $m = m_1 + m_2$ is a good quantum number on both sides of the transformation, the square matrix just considered may be split up into submatrices, each corresponding to a given value of m. Each submatrix is itself unitary, and so we have

$$(3.5.5) \qquad \sum_{j} (j_1\, m_1'\, j_2\, m-m_1' | j_1\, j_2\, j\, m)(j_1\, j_2\, j\, m | j_1\, m_1\, j_2\, m-m_1)$$
$$= \delta_{m_1' m_1}$$

$$(3.5.6) \qquad \sum_{m_1} (j_1\, j_2\, j'\, m | j_1\, m_1\, j_2\, m-m_1)(j_1\, m_1\, j_2\, m-m_1 | j_1\, j_2\, j\, m)$$
$$= \delta_{j' j} \delta(j_1\, j_2\, j)$$

RECURSION RELATIONS. We shall now derive recursion relations between the vector-coupling coefficients, which with the above equations (3.5.5) and (3.5.6) and the phase conventions (3.4.1) and (3.4.2) will enable us to determine completely all coefficients.

First we define the function $A(j, m)$ which will be used in the ensuing calculations. It is given by

$$(3.5.7) \qquad A(j, m) \equiv [(j+m)(j-m+1)]^{\frac{1}{2}}$$

Evidently

$$(3.5.8) \qquad A(j, m+1) = A(j, -m) = [(j+m+1)(j-m)]^{\frac{1}{2}}$$

and
$$A(j, -m+1) = A(j, m)$$

Now suppose we know all the coefficients $(j_1 \, m_1 \, j_2 \, m_2 | j_1 \, j_2 \, j \, m)$ for a given j and m, i.e. we know the state $w(\gamma \, j_1 \, j_2 \, j \, m)$ in terms of the $v(\gamma \, j_1 \, m_1 \, j_2 \, m_2)$. We may use the operator $J_- = J_{1-} + J_{2-}$ to obtain $w(\gamma \, j_1 \, j_2 \, j \, m-1)$, and shall express the result in the $m_1 m_2$ scheme. Since the quantum numbers j_1 and j_2 are unaltered in the following considerations, we drop them temporarily in the interest of clarity.

$$J_- \sum_{m_1 m_2} v(m_1 \, m_2)(m_1 \, m_2 | j \, m)$$
$$= (j \, m-1 | J_- | j \, m) \sum_{m_1' m_2'} v(m_1' \, m_2')(m_1' \, m_2' | j \, m-1)$$

This is equal to the result of the application of $J_{1-} + J_{2-}$:

$$\sum_{m_1 m_2} (j_1 \, m_1-1 | J_{1-} | j_1 \, m_1) v(m_1-1 \, m_2)(m_1 \, m_2 | j \, m)$$
$$+ \sum_{m_1' m_2'} (j_2 \, m_2'-1 | J_{2-} | j_2 \, m_2') v(m_1' \, m_2'-1)(m_1' \, m_2' | j \, m)$$

We substitute for the matrix elements of J_-, J_{1-}, and J_{2-} (see (2.3.17)) and equate coefficients of $v(m_1 \, m_2)$ in the two expressions above, making use of the $A(j, m)$ notation just defined.

(3.5.9) $\quad A(j, m)(m_1 \, m_2 | j \, m-1) = A(j_1, m_1+1)(m_1+1 \, m_2 | j \, m)$
$$+ A(j_2, m_2+1)(m_1 \, m_2+1 | j \, m)$$

A similar recursion relation is obtained by application of $J_+ = J_{1+} + J_{2+}$:

(3.5.10) $\quad A(j, m+1)(m_1 \, m_2 | j \, m+1) = A(j_1, m_1)(m_1-1 \, m_2 | j \, m)$
$$+ A(j_2, m_2)(m_1 \, m_2-1 | j \, m)$$

THE PHASES OF CERTAIN V-C COEFFICIENTS. The conventions (3.4.1) and (3.4.2) are now used to determine the argument of $(j_1 \, j_1 \, j_2 \, m_2 | j_1 \, j_2 \, j \, j)$ i.e. those coefficients where $m_1 = j_1$, $m = j$. We consider the matrix elements of the operator product $J_+ J_{1z}$. The commutation relations (3.2.4) show that

$$J_+ J_{1z} = -\hbar J_{1+} + J_{1z} J_+$$

The matrix component of this equation between states $j+1, j+1$; j, j is

$$(j+1 \, j+1 | J_+ J_{1z} | j \, j) = (j+1 \, j+1 | J_+ | j+1 \, j)(j+1 \, j | J_{1z} | j \, j)$$
$$= -\hbar(j+1 \, j+1 | J_{1+} | j \, j)$$

Note that there is only one nonzero term in the summation on the left and that the matrix element of $J_{1z}J_+$ is zero. Now the left-hand side is real and positive as a result of conventions (2.3.16) and (3.4.2). Hence

$$-\hbar \sum_{m_1 m_1'} (j_1 \, j_2 \, j+1 \, j+1 | j_1 \, m_1 \, j_2 \, m_2)(j_1 \, m_1 | J_{1+} | j_1 \, m_1')$$
$$\times (j_1 \, m_1' \, j_2 \, m_2 | j_1 \, j_2 \, j \, j) > 0$$

(Here and in the following work m_2 is assumed to take the value $m - m_1$ so that the V-C coefficient in question is nonzero.) The recursion relation (3.5.10) shows that when $m = j$, the sign of $(j_1 \, m_1 \, j_2 \, m_2 | j_1 \, j_2 \, j \, j)$ alternates with m_1. I.e.

$$\arg (j_1 \, m_1 \, j_2 \, m_2 | j_1 \, j_2 \, j \, j) = (-1)^{j_1 - m_1} \arg (j_1 \, j_1 \, j_2 \, m_2' | j_1 \, j_2 \, j \, j).$$

We have therefore from the above inequality that

$$\arg (j_1 \, j_1 \, j_2 \, m_2 | j_1 \, j_2 \, j \, j) \arg (j_1 \, j_1 \, j_2 \, m_2' | j_1 \, j_2 \, j+1 \, j+1)$$
$$\times \sum_{m_1 m_1'} |(j_1 \, m_1 \, j_2 \, m_2 | j_1 \, j_2 \, j \, j)|(j_1 \, m_1 | J_{1+} | j_1 \, m_1')$$
$$\times |(j_1 \, m_1' \, j_2 \, m_2' | j_1 \, j_2 \, j+1 \, j+1)| > 0$$

The matrix element of J_{1+} is real and positive by (2.3.16). Hence

$$\arg (j_1 \, j_1 \, j_2 \, m_2 | j_1 \, j_2 \, j \, j) \cdot \arg (j_1 \, j_1 \, j_2 \, m_2' | j_1 \, j_2 \, j+1 \, j+1) = 1$$

But we know from (3.4.1) that $\arg (j_1 \, j_1 \, j_2 \, j_2 | j_1 \, j_2 \, j_1+j_2 \, j_1+j_2) = 1$. Hence

(3.5.11) $$\arg (j_1 \, j_1 \, j_2 \, m_2 | j_1 \, j_2 \, j \, j) = 1$$

for all allowed j.

REALITY OF THE V-C COEFFICIENTS. All the V-C coefficients with given j_1, j_2, and j are connected by the recursion relations (3.5.9), (3.5.10). These relations have real coefficients, so the fact that we have shown in (3.5.11) that one of the set is real implies that they are all real. That is, the use of the conventions (3.4.1) and (3.4.2) results in the reality of all V-C coefficients.

CASE WHEN ONE OF THE j VALUES IS ZERO. The V-C coefficients are easily evaluated when this happens. We see from (3.4.1) that

(3.5.12) $$(j \, m \, 0 \, 0 | j \, 0 \, j \, m) = 1$$

When the resultant $j = 0$ and $j_1 = j_2$, the recursion relation (3.5.9) shows that the coefficients $(j_1 \, m_1 \, j_1 \, -m_1 | j_1 \, j_1 \, 0 \, 0)$ are independent of m_1 apart from the sign, which alternates with m_1. The unitary condition (3.5.6) and the result (3.5.11) show that

(3.5.13) $$(j_1 \, m_1 \, j_1 \, -m_1 | j_1 \, j_1 \, 0 \, 0) = (-1)^{j_1 - m_1}(2j_1 + 1)^{-\frac{1}{2}}$$

SYMMETRY PROPERTIES OF THE V-C COEFFICIENT. When we have to deal with the addition of two angular momenta, say $\mathbf{J}_a$ and $\mathbf{J}_b$, we must pay attention to the order in which they are coupled, i.e. which of the two is associated with the angular momentum $\mathbf{J}_1$ in the preceding arguments. The reason for this is apparent when we recall convention (3.4.2), namely that all matrix elements of J_{1z} nondiagonal in j are chosen to be non-negative, which implies that the corresponding matrix elements of J_{2z} are nonpositive.

It follows that the matrix elements of J_{az} in the schemes $(\gamma\, j_a\, j_b\, j\, m)$ and $(\gamma\, j_b\, j_a\, j\, m)$ are of opposite sign. Now J_{az} connects only those states whose j's differ by one or zero (cf. (3.3.3)); hence for successive values of j the eigenvectors $w(j_a\, j_b\, j\, m)$ and $w(j_b\, j_a\, j\, m)$ will change their relative phase. However, when $j = j_a + j_b$ the eigenvectors will have the same phase, for the convention (3.4.1) implies that the states $w(j_a\, j_b\, j_a+j_b\, j_a+j_b)$ and $w(j_b\, j_a\, j_b+j_a\, j_b+j_a)$ are identical and the states with other values of m may be obtained simply by application to these of the operator J_- a sufficient number of times (cf. (2.3.17)). Hence for a general value of j we must have

$$w(j_a\, j_b\, j\, m) = (-1)^{j_a+j_b-j} w(j_b\, j_a\, j\, m).$$

The V-C coefficients are related accordingly:

(3.5.14) $\quad (j_a\, m_a\, j_b\, m_b | j_a\, j_b\, j\, m) = (-1)^{j_a+j_b-j}(j_b\, m_b\, j_a\, m_a | j_b\, j_a\, j\, m)$

We may obtain other symmetry relations for the V-C coefficients by recourse to the concept of time reversal (2.8). We replace the operator equation

$$\mathbf{J}_1 + \mathbf{J}_2 = \mathbf{J}$$

by

$$\mathbf{J}_1 = -\mathbf{J}_2 + \mathbf{J}_3 = \tilde{\mathbf{J}}_2 + \mathbf{J}_3$$

where $\tilde{\mathbf{J}}_2 \equiv -\mathbf{J}_2$ is a "time reversed" angular momentum operator. This operator has eigenvectors $\tilde{u}(j_2\, m_2)$ which are related to those of $\mathbf{J}_2$ by (2.8.2).

These results suggest that the coefficient $(j_2\, -m_2\, j_3\, m_3 | j_2\, j_3\, j_1\, m_1)$ may be related to $(j_1\, m_1\, j_2\, m_2 | j_1\, j_2\, j_3\, m_3)$. We investigate this possibility by writing down the recursion relations for $(j_2\, -m_2\, j_3\, m_3 | j_2\, j_3\, j_1\, m_1)$ corresponding to (3.5.9) and (3.5.10); we make use of the symmetry properties of the function $A(j, m)$ (see (3.5.8))

$A(j_1, m_1)(j_2\, -m_2\, j_3\, m_3 | j_2\, j_3\, j_1\, m_1-1)$
$\quad = A(j_2, m_2)(j_2\, -m_2+1\, j_3\, m_3 | j_2\, j_3\, j_1\, m_1)$
$\quad\quad + A(j_3, m_3+1)(j_2\, -m_2\, j_3\, m_3+1 | j_2\, j_3\, j_1\, m_1)$

$A(j_1, m_1+1)(j_2\, -m_2\, j_3\, m_3 | j_2\, j_3\, j_1\, m_1+1)$
$\quad = A(j_2, m_2+1)(j_2\, -m_2-1\, j_3\, m_3 | j_2\, j_3\, j_1\, m_1)$
$\quad\quad + A(j_3, m_3)(j_2\, -m_2\, j_3\, m_3-1 | j_2\, j_3\, j_1\, m_1)$

On comparing these recursion relations with the original ones (3.5.9) and (3.5.10) we see that the quantity $(-1)^{m_2}(j_2 - m_2\, j_3\, m_3 | j_2\, j_3\, j_1\, m_1)$ has the same recursion relations as $(j_1\, m_1\, j_2\, m_2 | j_1\, j_2\, j_3\, m_3)$. It follows that these quantities differ only by a factor independent of the magnetic quantum numbers

$$(j_1\, m_1\, j_2\, m_2 | j_1\, j_2\, j_3\, m_3) = C \cdot (-1)^{m_2}(j_2 - m_2\, j_3\, m_3 | j_2\, j_3\, j_1\, m_1)$$

The modulus of C is easily found by use of the unitary property (3.5.6). The argument of C is given by (3.5.11) to be $(-1)^{j_2}$. For we have, taking the special values $m_1 = j_1$, $m_3 = j_3$,

$$\arg\,(j_1\, j_1\, j_2\, j_3 - j_1 | j_1\, j_2\, j_3\, j_3) = 1;$$

$$\arg\,(j_2\, j_1 - j_3\, j_3\, j_3 | j_2\, j_3\, j_1\, j_1) = (-1)^{j_2 + j_3 - j_1} \quad \text{(by (3.5.11) and (3.5.14))}$$

The final symmetry relation is

(3.5.15)
$$(j_1\, m_1\, j_2\, m_2 | j_1\, j_2\, j_3\, m_3)$$
$$= (-1)^{j_2 + m_2}\left(\frac{2j_3 + 1}{2j_1 + 1}\right)^{\frac{1}{2}}(j_2 - m_2\, j_3\, m_3 | j_2\, j_3\, j_1\, m_1)$$

Other symmetry relations of this type are obtained in the same way. For example

(3.5.16)
$$(j_1\, m_1\, j_2\, m_2 | j_1\, j_2\, j_3\, m_3)$$
$$= (-1)^{j_1 - m_1}\left(\frac{2j_3 + 1}{2j_2 + 1}\right)^{\frac{1}{2}}(j_3\, m_3\, j_1 - m_1 | j_3\, j_1\, j_2\, m_2)$$

We may reverse the signs of all three m's by applying (3.5.15) three times. This gives the relation

(3.5.17)
$$(j_1\, m_1\, j_2\, m_2 | j_1\, j_2\, j_3\, m_3)$$
$$= (-1)^{j_1 + j_2 - j_3}(j_1 - m_1\, j_2 - m_2 | j_1\, j_2\, j_3 - m_3)$$

3.6. Computation of the Vector-Coupling Coefficients

The problem in hand is the computation of the matrix of vector-coupling coefficients belonging to a given pair of values of j_1 and j_2. We have already remarked that this matrix may be split into unitary submatrices corresponding to the possible values of m. The elements of these submatrices are linked by the recursion relations (3.5.9) and (3.5.10), and the submatrix for $m = j_1 + j_2$, with one element, is specified by the convention (3.4.1).

We shall see how, by use of the recursion relation (3.5.9), all coefficients with a given j may be computed from those with the maximum m value, namely $m = j$. The latter coefficients are obtained by use of the recursion

3.6 · COMPUTATION OF THE COEFFICIENTS

relation (3.5.10), the unitary condition (3.5.6), and the phase convention (3.4.1).

Thus all coefficients of the form $(j_1 \, m_1 \, j_2 \, m_2 | j_1 \, j_2 \, j \, j)$ are computed first. For the sake of clarity we shall omit the arguments j_1, j_2 in the symbols representing the V-C coefficients, which are not directly relevant to the calculation in hand.

We specialize the recursion relation (3.5.10) to give

$$(3.6.1) \quad 0 = [(j_1 + m_1)(j_1 - m_1 + 1)]^{\frac{1}{2}} (m_1 - 1 \, m_2 | j \, j)$$
$$+ [(j_2 + j - m_1 + 1)(j_2 - j + m_1)]^{\frac{1}{2}} (m_1 \, m_2' | j \, j)$$

I.e.

$$(m_1 - 1 \, m_2 | j \, j) = - \left[\frac{(j_2 + j - m_1 + 1)(j_2 - j + m_1)}{(j_1 + m_1)(j_1 - m_1 + 1)} \right]^{\frac{1}{2}} (m_1 \, m_2' | j \, j)$$

which by successive application gives

$$(m_1 \, m_2 | j \, j)$$
$$(3.6.2) \quad = (-1)^{j_1 - m_1} \left[\frac{(j_2 + j - m_1)!(j_1 + j_2 - j)!(j_1 + m_1)!}{(2j_1)!(-j_1 + j_2 + j)!(j_2 - j + m_1)!(j_1 - m_1)!} \right]^{\frac{1}{2}}$$
$$\times (j_1 \, m_2' | j \, j)$$

The magnitude of $(j_1 \, j_1 \, j_2 \, j - j_1 | j_1 \, j_2 \, j \, j)$ is obtained by use of the unitary condition

$$(3.6.3) \quad \sum_{m_1 = -j_1}^{j_1} |(m_1 \, m_2 | j \, j)|^2 = 1$$

That is, we have

$$|(j_1 \, m_2 | j \, j)|^2 \sum_{m_1} \frac{(j_2 + j - m_1)!(j_1 + j_2 - j)!(j_1 + m_1)!}{(2j_1)!(-j_1 + j_2 + j)!(j_2 - j + m_1)!(j_1 - m_1)!} = 1$$

Now equation A.1.3 in Appendix 1 gives the sum over m_1,

$$(3.6.4) \quad \sum_{m_1} \frac{(j_1 + m_1)!(j_2 + j - m_1)!}{(j_1 - m_1)!(j_2 - j + m_1)!}$$
$$= \frac{(j_1 + j_2 + j + 1)!(-j_1 + j_2 + j)!(j_1 - j_2 + j)!}{(2j + 1)!(j_1 + j_2 - j)!}$$

Hence

$$(3.6.5) \quad |(j_1 \, j_1 \, j_2 \, j - j_1 | j_1 \, j_2 \, j \, j)| = \left[\frac{(2j_1)!(2j + 1)!}{(j_1 + j_2 + j + 1)!(j_1 - j_2 + j)!} \right]^{\frac{1}{2}}$$

The phase of this coefficient is given by (3.5.11):

$$(3.6.6) \quad (j_1 \, j_1 \, j_2 \, j - j_1 | j_1 \, j_2 \, j \, j) = + \left[\frac{(2j_1)!(2j + 1)!}{(j_1 + j_2 + j + 1)!(j_1 - j_2 + j)!} \right]^{\frac{1}{2}}$$

We now go on to compute the general V-C coefficient from those with $m = j$. The recursion relation (3.5.9) is rewritten as

$$[(j + m)(j - m + 1)]^{\frac{1}{2}}(m_1\ m_2|j\ m-1)$$
(3.6.7)
$$= [(j_1 - m_1)(j_1 + m_1 + 1)]^{\frac{1}{2}}(m_1+1\ m_2|j\ m)$$
$$+ [(j_2 - m + m_1 + 1)(j_2 + m - m_1)]^{\frac{1}{2}}(m_1\ m_2+1|j\ m)$$

We may express this relation as

(3.6.8) $\quad q(m_1,\ m-1)(m_1\ m_2|j\ m-1)$
$$= q(m_1+1,\ m)(m_1+1\ m_2|j\ m) - q(m_1,\ m)(m_1\ m_2+1|j\ m)$$

where

(3.6.9) $\quad q(m_1,\ m) = (-1)^{m+m_1}\left[\dfrac{(j_1 + m_1)!(j_2 + m - m_1)!(j - m)!}{(j_1 - m_1)!(j_2 - m + m_1)!(j + m)!}\right]^{\frac{1}{2}}$

Making use of the finite difference notation, we have

$$q(m_1,\ m-1)(m_1\ m_2|j\ m-1) = \underset{m_1}{\Delta}\{q(m_1,\ m)(m_1\ m_2'|j\ m)\}$$

and

$$q(m_1,\ m)(m_1\ m_2|j\ m) = \underset{m_1}{\Delta^{j-m}}\{q(m_1,\ j)(m_1\ j-m_1|j\ j)\}$$

Now the nth difference of a function $f(x)$ is given by[1]

$$\underset{x}{\Delta^n} f(x) = \sum_{\nu=0}^{n}(-1)^{n+\nu}\binom{n}{\nu}f(x+\nu)$$

Therefore

$(j_1\ m_1\ j_2\ m_2|j_1\ j_2\ j\ m)$

(3.6.10)
$$= \dfrac{(-1)^{j-m}}{q(m_1,\ m)}\sum_{s=0}^{j-m}(-1)^s\binom{j-m}{s}q(m_1+s,\ j)(m_1+s\ m_2'|j\ j)$$

$$= \delta(m_1+m_2,\ m)\left[\dfrac{(2j+1)(j_1+j_2-j)!(j_1-m_1)!(j_2-m_2)!(j+m)!(j-m)!}{(j_1+j_2+j+1)!(j_1-j_2+j)!(-j_1+j_2+j)!(j_1+m_1)!(j_2+m_2)!}\right.$$

$$\times \sum_s(-1)^{s+j_1-m_1}\dfrac{(j_1+m_1+s)!(j_2+j-m_1-s)!}{s!(j_1-m_1-s)!(j-m-s)!(j_2-j+m_1+\ }$$

where we have substituted from (3.6.2), (3.6.6) and (3.6.9) and the summation is over positive integer s such that the arguments in the denominator are non-negative. This formula is identical with that obtained by Racah (1942, eq. 15), and we follow his method for transforming it into a more symmetric expression, by making use of equations A.1.1 and A.1.2 in Appendix 1.

---

[1] Cf. Jordan (1947), Milne-Thomson (1933), etc.

These give

$$\sum_s (-1)^{s+j_1-m_1} \frac{(j_1+m_1+s)!(j_2+j-m_1-s)!}{s!(j_1-m_1-s)!(j-m-s)!(j_2-j+m_1+s)!}$$

$$= \sum_{su} (-1)^{s+j_1-m_1} \frac{(j_1+m_1+s)!}{s!(j_2-j+m_1+s)!} \frac{(j_2+m_2)!(-j_1+j_2+j)!}{(j_2+m_2-u)!(-j_1+j_2+j-u)!(j_1-j_2-m-s+u)!u!}$$

$$= \sum_u (-1)^{j_2+m_2-u} \frac{(j_1+m_1)!(j_1-j_2+j)!(j_2+m_2)!(-j_1+j_2+j)!}{(j_1-j_2-m+u)!(j_1-j-m_2+u)!(j+m-u)!(j_2+m_2-u)!(-j_1+j_2+j-u)!u!}$$

On putting $z = j_2 + m_2 - u$ we have

$$(j_1\ m_1\ j_2\ m_2|j_1\ j_2\ j\ m) = \delta(m_1+m_2,m)\left[\frac{(2j+1)(j_1+j_2-j)!(j_1-j_2+j)!(-j_1+j_2+j)!}{(j_1+j_2+j+1)!}\right]^{\frac{1}{2}}$$

(3.6.11)

$$\times\ [(j_1+m_1)!(j_1-m_1)!(j_2+m_2)!(j_2-m_2)!(j+m)!(j-m)!]^{\frac{1}{2}}$$

$$\times \sum_z (-1)^z \frac{1}{z!(j_1+j_2-j-z)!(j_1-m_1-z)!(j_2+m_2-z)!(j-j_2+m_1+z)!(j-j_1-m_2+z)!}$$

A number of derivations of the general formula for the vector coupling coefficient have been given. That of Racah (1942) has already been mentioned; other notable derivations are that of Wigner (1931), which makes use of group theoretical methods, and that of Schwinger (1952) in which an elegant operator method is employed.[2]

We may obtain from (3.6.11) simpler formulas for certain values of the arguments of the V-C coefficient.

(3.6.12)
$$(j_1\ m_1\ j_2\ m_2|j_1\ j_2\ j_1+j_2\ m_1+m_2) = \left[\frac{(2j_1)!(2j_2)!(j_1+j_2+m_1+m_2)!(j_1+j_2-m_1-m_2)!}{(2j_1+2j_2)!(j_1-m_1)!(j_1+m_1)!(j_2-m_2)!(j_2+m_2)!}\right]^{\frac{1}{2}}$$

(3.6.13)
$$(j_1\ j_1\ j_2\ m-j_1|j_1\ j_2\ j\ m) = \left[\frac{(2j+1)(2j_1)!(-j_1+j_2+j)!(j_1+j_2-m)!(j+m)!}{(j_1+j_2-j)!(j_1-j_2+j)!(j_1+j_2+j+1)!(-j_1+j_2+m)!(j-m)!}\right]^{\frac{1}{2}}$$

3.7. The Wigner 3-j Symbol

INTRODUCTION OF THE CONCEPT OF CONTRAGREDIENT QUANTITIES. Let us consider the coupling of two angular momentum eigenvectors with the same j to form a state with zero angular momentum. This gives (see (3.5.13))

$$\sum_m u_1(j\ m)u_2(j\ -m)(-1)^{j-m} = (2j+1)^{\frac{1}{2}}v(j\ j\ 0\ 0)$$

---

[2] I have obtained this symmetric expression (3.6.11) directly and rapidly by an adaptation of the symbolic method of Kramers (cf. Kramers (1930), (1931); Brinkman (1956)); however, the approach is rather different from that otherwise used in this book.

Since the right-hand side is invariant under rotations, we may say that the quantities $(-1)^{i-m} u(j\ -m)$ transform under rotations *contragrediently*[3] to the $u(j\ m)$. We may also, following Wigner, introduce a quantity which behaves like a metric tensor, namely

$$(3.7.1) \qquad \begin{pmatrix} j \\ m\ m' \end{pmatrix} \equiv (-1)^{i+m} \delta_{m,\,-m'}.$$

That is, we have

$$(3.7.2) \qquad \sum_{mm'} u(j\ m) u(j\ m') \begin{pmatrix} j \\ m\ m' \end{pmatrix} = \text{invariant}$$

These properties of the angular momentum eigenvectors are closely associated with their behaviour under time reversal[4] and under rotation of the frame of reference through 180° about the y axis. These matters are discussed in Chapters 2 and 4.

The concept of contragredient quantities leads us to the conclusion that the vector-coupling coefficients are components of mixed tensors, thus giving some explanation of their unsymmetric properties (cf. (3.5.14), (3.5.15), (3.5.17), etc.)

A more symmetric quantity may thus be found by carrying out an operation corresponding to raising or lowering of indices in tensor algebra. Such a result is obtained by considering not the coefficient associated with coupling j_1 and j_2 to give j_3, but with the coupling of three angular momenta j_1, j_2, and j_3 to a resultant zero. However the phase of the resulting quantity is important, since it is of advantage to have maximum symmetry.

DEFINITION OF THE 3-j SYMBOL. This maximum symmetry is obtained in the so-called 3-j symbol of Wigner (1951) which is defined by

$$(3.7.3) \qquad \begin{pmatrix} j_1 & j_2 & j_3 \\ m_1 & m_2 & m_3 \end{pmatrix} = (-1)^{j_1 - j_2 - m_3} (2j_3 + 1)^{-\frac{1}{2}} (j_1\ m_1\ j_2\ m_2 | j_1\ j_2\ j_3\ -m_3)$$

Its symmetry properties are easily derived from those of the V-C coefficient. We have that an *even* permutation of the columns leaves the numerical value unchanged:

$$(3.7.4) \qquad \begin{pmatrix} j_1 & j_2 & j_3 \\ m_1 & m_2 & m_3 \end{pmatrix} = \begin{pmatrix} j_2 & j_3 & j_1 \\ m_2 & m_3 & m_1 \end{pmatrix} = \begin{pmatrix} j_3 & j_1 & j_2 \\ m_3 & m_1 & m_2 \end{pmatrix}$$

---

[3] See Weyl (1931) Chap. I, §3.
[4] Cf. (2.8) on "time reversed" eigenvectors.

3.7 · THE WIGNER 3-j SYMBOL

while an *odd* permutation is equivalent to multiplication by $(-1)^{j_1+j_2+j_3}$:

(3.7.5)
$$(-1)^{j_1+j_2+j_3}\begin{pmatrix} j_1 & j_2 & j_3 \\ m_1 & m_2 & m_3 \end{pmatrix} = \begin{pmatrix} j_2 & j_1 & j_3 \\ m_2 & m_1 & m_3 \end{pmatrix}$$
$$= \begin{pmatrix} j_1 & j_3 & j_2 \\ m_1 & m_3 & m_2 \end{pmatrix} = \begin{pmatrix} j_3 & j_2 & j_1 \\ m_3 & m_2 & m_1 \end{pmatrix}$$

The analogue of (3.5.17) is

(3.7.6)
$$\begin{pmatrix} j_1 & j_2 & j_3 \\ m_1 & m_2 & m_3 \end{pmatrix} = (-1)^{j_1+j_2+j_3}\begin{pmatrix} j_1 & j_2 & j_3 \\ -m_1 & -m_2 & -m_3 \end{pmatrix}$$

These symmetry properties should be compared with those of the similar symmetrized coefficients of Racah, Fano, etc. (cf. Table 3.1 at the end of this chapter).

It may be seen from the symmetry properties that certain 3-j symbols must be identically zero. In this class for example, are

$$\begin{pmatrix} \tfrac{3}{2} & \tfrac{3}{2} & 2 \\ \tfrac{1}{2} & \tfrac{1}{2} & -1 \end{pmatrix}, \quad \begin{pmatrix} 2 & 2 & 3 \\ 1 & 1 & -2 \end{pmatrix}$$

and any 3-j symbol with $m_1 = m_2 = m_3 = 0$, and $j_1 + j_2 + j_3$ odd.

The orthogonality properties are not so convenient. They are

(3.7.7)
$$\sum_{j_3 m_3} (2j_3 + 1)\begin{pmatrix} j_1 & j_2 & j_3 \\ m_1 & m_2 & m_3 \end{pmatrix}\begin{pmatrix} j_1 & j_2 & j_3 \\ m_1' & m_2' & m_3 \end{pmatrix} = \delta_{m_1 m_1'}\delta_{m_2 m_2'}$$

(3.7.8)
$$\sum_{m_1 m_2}\begin{pmatrix} j_1 & j_2 & j_3 \\ m_1 & m_2 & m_3 \end{pmatrix}\begin{pmatrix} j_1 & j_2 & j_3' \\ m_1 & m_2 & m_3' \end{pmatrix}$$
$$= (2j_3 + 1)^{-1}\delta_{j_3 j_3'}\delta_{m_3 m_3'}\delta(j_1 j_2 j_3)$$

where $\delta(j_1 j_2 j_3) = 1$ if j_1, j_2, j_3 satisfy the triangular condition, and is zero otherwise.

The greater symmetry of the 3-j symbol will be found useful when it is necessary to evaluate such quantities numerically; however the notation finds its main application in the discussion of the properties of the 6-j and 9-j symbols which, as is explained in Chapter 6, are invariant quantities built up from vector-coupling coefficients.

SPECIALIZED FORMULAS FOR THE 3-j SYMBOL. The formulas given in (3.6) for certain values of the arguments of the V-C coefficients are repeated here for the 3-j symbol

(3.7.9) $\begin{pmatrix} j & j & 0 \\ m & -m & 0 \end{pmatrix} = (-1)^{j-m}(2j+1)^{-\frac{1}{2}}$

(3.7.10) $\begin{pmatrix} j_1 & j_2 & j_1+j_2 \\ m_1 & m_2 & -m_1-m_2 \end{pmatrix} = (-1)^{j_1-j_2+m_1+m_2}\left[\frac{(2j_1)!(2j_2)!(j_1+j_2+m_1+m_2)!(j_1+j_2-m_1-m_2)!}{(2j_1+2j_2+1)!(j_1+m_1)!(j_1-m_1)!(j_2+m_2)!(j_2-m_2)!}\right]^{\frac{1}{2}}$

(3.7.11) $\begin{pmatrix} j_1 & j_2 & j_3 \\ j_1 & -j_1-m_3 & m_3 \end{pmatrix}$

$= (-1)^{-j_1+j_2+m_3}\left[\frac{(2j_1)!(-j_1+j_2+j_3)!(j_1+j_2+m_3)!(j_3-m_3)!}{(j_1+j_2+j_3+1)!(j_1-j_2+j_3)!(j_1+j_2-j_3)!(-j_1+j_2-m_3)!(j_3+m_3)!}\right]^{\frac{1}{2}}$

RECURSION RELATIONS FOR THE 3-j SYMBOL. A number of useful recursion relations for the 3-j symbols may be obtained from an expression (6.2.8) for a product of a 3-j symbol and a 6-j symbol which is given in Chapter 6. These relations are got by giving special values to the arguments l_1, l_2, and l_3 of the 6-j symbol and evaluating the 6-j symbol and some of the 3-j symbols by use of Tables 5 and 2.

Let us take for example, $l_1 = \frac{1}{2}$, $l_2 = j_3 - \frac{1}{2}$, $l_3 = j_2 - \frac{1}{2}$. The sum on the right reduces to two terms, and we obtain finally

$$[(J+1)(J-2j_1)]^{\frac{1}{2}}\begin{pmatrix} j_1 & j_2 & j_3 \\ m_1 & m_2 & m_3 \end{pmatrix}$$

(3.7.12) $= [(j_2+m_2)(j_3-m_3)]^{\frac{1}{2}}\begin{pmatrix} j_1 & j_2-\frac{1}{2} & j_3-\frac{1}{2} \\ m_1 & m_2-\frac{1}{2} & m_3+\frac{1}{2} \end{pmatrix}$

$- [(j_2-m_2)(j_3+m_3)]^{\frac{1}{2}}\begin{pmatrix} j_1 & j_2-\frac{1}{2} & j_3-\frac{1}{2} \\ m_1 & m_2+\frac{1}{2} & m_3-\frac{1}{2} \end{pmatrix}$

where $J = j_1 + j_2 + j_3$.

Alternatively, we may take $l_1 = 1$, $l_2 = j_3 - 1$, $l_3 = j_2$. The recursion relation obtained is now

$$[(J+1)(J-2j_1)(J-2j_2)(J-2j_3+1)]^{\frac{1}{2}}\begin{pmatrix} j_1 & j_2 & j_3 \\ m_1 & m_2 & m_3 \end{pmatrix}$$

$= [(j_2-m_2)(j_2+m_2+1)(j_3+m_3)(j_3+m_3-1)]^{\frac{1}{2}}$

$\times \begin{pmatrix} j_1 & j_2 & j_3-1 \\ m_1 & m_2+1 & m_3-1 \end{pmatrix}$

(3.7.13) $- 2m_2[(j_3+m_3)(j_3-m_3)]^{\frac{1}{2}}\begin{pmatrix} j_1 & j_2 & j_3-1 \\ m_1 & m_2 & m_3 \end{pmatrix}$

$- [(j_2+m_2)(j_2-m_2+1)(j_3-m_3)(j_3-m_3-1)]^{\frac{1}{2}}$

$\times \begin{pmatrix} j_1 & j_2 & j_3-1 \\ m_1 & m_2-1 & m_3+1 \end{pmatrix}$

where again $J = j_1 + j_2 + j_3$.

3.7 · THE WIGNER 3-j SYMBOL

Such recursion relations make it possible in principle to compute any 3-j symbols starting from the formulas in Table 2 on page 125.

COMPUTATION OF 3-j SYMBOLS WITH $m_1 = m_2 = m_3 = 0$. We may use the recursion relation (3.7.13) to get the general formula for the frequently occurring symbol

$$\begin{pmatrix} j_1 & j_2 & j_3 \\ 0 & 0 & 0 \end{pmatrix}$$

We have first

(3.7.14) $\quad \begin{pmatrix} j_1 & j_2 & j_3 \\ 0 & 0 & 0 \end{pmatrix} = 0 \quad \text{if} \quad j_1 + j_2 + j_3 \text{ is odd}$

This is a consequence of the symmetry (3.7.6) of the 3-j symbol. If $J = j_1 + j_2 + j_3$ is even, we have from (3.7.13),

(3.7.15)
$$\begin{pmatrix} j_1 & j_2 & j_3 \\ 0 & 0 & 0 \end{pmatrix} = 2 \left[\frac{j_2(j_2+1)(j_3-1)j_3}{(J+1)(J-2j_1)(J-2j_2)(J-2j_3+1)} \right]^{\frac{1}{2}}$$
$$\times \begin{pmatrix} j_1 & j_2 & j_3-1 \\ 0 & +1 & -1 \end{pmatrix}$$

On applying (3.7.13) again,

(3.7.16)
$$\begin{pmatrix} j_1 & j_2 & j_3 \\ 0 & 0 & 0 \end{pmatrix} = \left[\frac{(J-2j_2-1)(J-2j_3+2)}{(J-2j_2)(J-2j_3+1)} \right]^{\frac{1}{2}}$$
$$\times \begin{pmatrix} j_1 & j_2+1 & j_3-1 \\ 0 & 0 & 0 \end{pmatrix}$$

The ν-fold iteration of this relation implies

$$\begin{pmatrix} j_1 & j_2 & j_3 \\ 0 & 0 & 0 \end{pmatrix} = \left[\frac{(J-2j_2)!(J-2j_3)!}{(J-2j_2-2\nu)!(J-2j_3+2\nu)!} \right]^{\frac{1}{2}}$$
$$\times \frac{\left(\frac{J}{2}-j_2-\nu\right)!\left(\frac{J}{2}-j_3+\nu\right)!}{\left(\frac{J}{2}-j_2\right)!\left(\frac{J}{2}-j_3\right)!} \begin{pmatrix} j_1 & j_2+\nu & j_3-\nu \\ 0 & 0 & 0 \end{pmatrix}$$

If we set $2\nu = J - 2j_2$ we have $j_2 + \nu = J/2$ and $j_3 - \nu = J/2 - j_1$. Thus (3.7.10) may be used to give

$$\begin{pmatrix} j_1 & \frac{J}{2} & \frac{J}{2}-j_1 \\ 0 & 0 & 0 \end{pmatrix} = \frac{(-1)^{J/2}\left(\frac{J}{2}\right)!}{j_1!\left(\frac{J}{2}-j_1\right)!} \left[\frac{(2j_1)!(J-2j_1)!}{(J+1)!} \right]^{\frac{1}{2}}$$

and we get finally

$$\begin{pmatrix} j_1 & j_2 & j_3 \\ 0 & 0 & 0 \end{pmatrix} = (-1)^{J/2} \left[\frac{(J-2j_1)!(J-2j_2)!(J-2j_3)!}{(J+1)!} \right]^{\frac{1}{2}}$$

(3.7.17)

$$\times \frac{\left(\frac{J}{2}\right)!}{\left(\frac{J}{2}-j_1\right)!\left(\frac{J}{2}-j_2\right)!\left(\frac{J}{2}-j_3\right)!}$$

if J is even.

3.8. Tabulation of Formulas and Numerical Values for Vector-Coupling Coefficients

It is not usually very practicable to obtain numerical values of the V-C coefficients from the general formula (3.6.11), and tables of formulas are available where one of the j values is fixed and the numerical value is given in terms of the remaining arguments. Such tables are given by Condon and Shortley (1935) for $j_2 = \frac{1}{2}, 1, \frac{3}{2}$ or 2. A similar table for $j = 3$ is given by Falkoff, et al. (1952), and one for $j = \frac{5}{2}$ by Saito and Morita (1955)

Tabulation of corresponding formulas for the 3-j symbols makes it easier to take advantage of the symmetry properties of these quantities; formulas for $j_3 = \frac{1}{2}, 1, \frac{3}{2}$, and 2 appear in Table 2.

Numerical values of a few V-C coefficients are given by Alder (1952); the most extensive tabulation to date is that of Simon (1954), who gives numerical values to ten decimal places of all V-C coefficients up to and including $j_1 = 4, j_2 = \frac{9}{2}, j = \frac{9}{2}$. Numerical values for a number of coefficients which differ only trivially from the 3-j symbols

$$\begin{pmatrix} j_1 & j_2 & j_3 \\ 0 & 0 & 0 \end{pmatrix}$$

have been given by Shortley and Fried (1938) and Sharp, et al. (1954).

COMPUTATION OF THE $\begin{pmatrix} j_1 & j_2 & j_3 \\ 0 & 0 & 0 \end{pmatrix}$. The quickest method is not to use the general formula (3.7.17) but to start from a

$$\begin{pmatrix} j & j & 0 \\ 0 & 0 & 0 \end{pmatrix}$$

given by (3.7.9) and to use the recursion relation (3.7.16) as many times as necessary. For example let us compute

$$\begin{pmatrix} 3 & 2 & 3 \\ 0 & 0 & 0 \end{pmatrix}$$

3.9 · TIME REVERSAL

We have from (3.7.9) that

$$\begin{pmatrix} 4 & 4 & 0 \\ 0 & 0 & 0 \end{pmatrix} = \begin{pmatrix} 4 & 0 & 4 \\ 0 & 0 & 0 \end{pmatrix} = \frac{1}{3}$$

The symmetry properties (3.7.4) and (3.7.5), together with (3.7.16) give

$$\left[\frac{8.1}{7.2}\right]^{\frac{1}{2}} \begin{pmatrix} 4 & 0 & 4 \\ 0 & 0 & 0 \end{pmatrix} = \begin{pmatrix} 4 & 1 & 3 \\ 0 & 0 & 0 \end{pmatrix} = \begin{pmatrix} 3 & 1 & 4 \\ 0 & 0 & 0 \end{pmatrix};$$

$$\left[\frac{6.1}{5.2}\right]^{\frac{1}{2}} \begin{pmatrix} 3 & 1 & 4 \\ 0 & 0 & 0 \end{pmatrix} = \begin{pmatrix} 3 & 2 & 3 \\ 0 & 0 & 0 \end{pmatrix}$$

therefore

$$\begin{pmatrix} 3 & 2 & 3 \\ 0 & 0 & 0 \end{pmatrix} = \left(\frac{4}{105}\right)^{\frac{1}{2}}$$

The coefficients of type

$$\begin{pmatrix} j_1 & j_2 & j_3 \\ 0 & 1 & -1 \end{pmatrix}$$

which are often used in angular correlation calculations may be got from the

$$\begin{pmatrix} j_1 & j_2 & j_3 \\ 0 & 0 & 0 \end{pmatrix}$$

by application of (3.7.15).

3.9. Time Reversal and the Eigenvectors Resulting from Vector Coupling

It is of interest from several points of view (see for example (5.5) and (5.11)) to study the properties under time reversal of the angular momentum eigenvectors resulting from vector coupling according to (3.5.1). In the case of integer j, as has been remarked in (2.8), time reversal is equivalent to the taking of the complex conjugate of the function concerned.

Let us first consider an angular momentum eigenvector resulting from coupling two systems represented by eigenvectors of the type $\mathfrak{Y}_{lm}$ which have the property (2.5.8) under complex conjugation:

$$\Psi(l_1\ l_2\ lm) = \sum_{m_1 m_2} (l_1\ l_2\ lm | l_1\ m_1\ l_2\ m_2) \mathfrak{Y}_{l_1 m_1} \mathfrak{Y}_{l_2 m_2}$$

The reality and symmetry (3.5.17) of the V-C coefficients shows us that

$\Psi(l_1 \, l_2 \, lm)$ has the same property under complex conjugation as the original $\mathfrak{Y}$'s:

(3.9.1) $\quad\quad \Psi^*(l_1 \, l_2 \, lm) = (-1)^{l+m}\Psi(l_1 \, l_2 \, l-m)$

On the other hand if we chose to use the Y_{lm} with the property (2.5.6) we should find that the resultant $\Psi(l_1 \, l_2 \, l \, m)$ would *not* share that property under complex conjugation.

The result (3.9.1) is generalized immediately to the case of general j's and time reversal; the angular momentum eigenvector obtained by vector coupling two sets of eigenvectors with the property (2.8.2) under time reversal itself has this property.

Table 3.1.

1. *Unsymmetrized V-C Coefficients.* These are all numerically equal (insofar as the authors have stated their assumptions about phases) to the V-C coefficient defined by Condon and Shortley. The symbols used for the angular momentum quantum numbers are the same throughout for ease of comparison, and in some places are different from those used by the authors mentioned.

   a. Biedenharn (1952) $\quad\quad C^{j_1 j_2 j}_{m_1 m_2 m}$
   b. Blatt and Weisskopf (1952) $\quad C_{j_1 j_2}(jm; m_1 m_2)$
   c. Condon and Shortley (1935), et al. $\quad \begin{Bmatrix} (j_1 \, j_2 \, j \, m | j_1 \, j_2 \, m_1 \, m_2) \\ (j_1 \, j_2 \, j \, m | j_1 \, m_1 \, j_2 \, m_2) \\ (j \, m | m_1 \, m_2) \end{Bmatrix}$ also $\begin{Bmatrix} (j_1 j_2 m_1 m_2 | j_1 j_2 jm) \\ \text{etc.} \end{Bmatrix}$
   d. Eckart (1930) $\quad\quad A^{j j_1 j_2}_{m m_1 m_2}$
   e. Fano (1952) $\quad\quad \langle j_1 \, m_1, j_2 \, m_2 | (j_1 \, j_2) j \, m \rangle$
   f. Jahn (1951), Alder (1952) $\quad C^{jm}_{j_1 m_1 j_2 m_2}$
   g. Rose (1953) $\quad\quad C(j_1 \, j_2 \, j; m_1 \, m_2)$
   h. van der Waerden (1931) Landau and Lifschitz (1948) $\quad C^j_{m_1 m_2}$
   i. Wigner (1931) $\quad\quad S^{j_1 j_2}_{j m_1 m_2}$
   j. Boys (1951) $\quad\quad X(j, m, j_1, j_2, m_1)$

2. *Symmetrized V-C Coefficients.* These are given relative to Wigner's 3-j symbol, which is given in terms of the V-C coefficient by (3.7.3).

   a. Fano (1952) $\quad\quad \langle j_1 m_1, j_2 m_2, j_3 m_3 | 0 \rangle = (-1)^{j_1 - j_2 + j_3} \begin{pmatrix} j_1 & j_2 & j_3 \\ m_1 & m_2 & m_3 \end{pmatrix}$

   b. Landau and Lifschitz (1948) $\quad S_{j_1 m_1; j_2 m_2; j_3 m_3} = (-1)^{j_1 - j_2 + j_3} \begin{pmatrix} j_1 & j_2 & j_3 \\ m_1 & m_2 & m_3 \end{pmatrix}$

   c. Racah (1942) $\quad\quad V(j_1 j_2 j_3; m_1 m_2 m_3) = (-1)^{j_1 + j_2 - j_3} \begin{pmatrix} j_1 & j_2 & j_3 \\ m_1 & m_2 & m_3 \end{pmatrix}$

   d. Schwinger (1952) $\quad\quad X(j_1 j_2 j_3; m_1 m_2 m_3) = \begin{pmatrix} j_1 & j_2 & j_3 \\ m_1 & m_2 & m_3 \end{pmatrix}$

CHAPTER 4

The Representations of Finite Rotations

4.1. The Transformations of the Angular Momentum Eigenvectors under Finite Rotations*

INTRODUCTORY REMARKS. We have seen in Chapter 2 how the $2j + 1$ angular momentum eigenvectors $u(j, -j)$, $u(j, -j+1)$, ..., $u(j, j)$ form a basis for an irreducible representation of the angular momentum operators. These operators are, when they are defined by expressions such as (2.1.3) and (2.1.4), proportional to the infinitesimal rotations. We may obtain *finite* rotations by iteration of the infinitesimal rotations, and hence the $2j + 1$ eigenvectors mentioned above form a basis for a representation of the finite rotations. In other words, under a finite rotation of the frame of reference a $u(jm)$ is transformed into a state vector which is an eigenvector of $\mathbf{J}^2$ with the same j. This is a familiar fact for the integer representations, for the $u(lm)$ are then in the r representation the spherical harmonics $Y_{lm}(\theta, \varphi)$; and although the angular momentum operators in the half-odd integer representations may not be expressed as differential operators in configuration space, we shall see that, in a certain sense, representations of finite rotations are given by the corresponding transformations of the $u(jm)$.

In the following discussions the term rotation will be interpreted as a rotation of the frame of reference about the origin, the field points (i.e. the physical system) being supposed fixed. Each point of three-dimensional space is thus given new coordinates, which are functions of the old coordinates and of the parameters which describe the rotation, namely the Euler angles.

The rotations 1), 2) and 3) (see p. 7) correspond to successive applications to the coordinate column vector of the three matrices:

$$(4.1.1) \quad \begin{bmatrix} \cos\gamma & \sin\gamma & 0 \\ -\sin\gamma & \cos\gamma & 0 \\ 0 & 0 & 1 \end{bmatrix} \begin{bmatrix} \cos\beta & 0 & -\sin\beta \\ 0 & 1 & 0 \\ \sin\beta & 0 & \cos\beta \end{bmatrix} \begin{bmatrix} \cos\alpha & \sin\alpha & 0 \\ -\sin\alpha & \cos\alpha & 0 \\ 0 & 0 & 1 \end{bmatrix} \begin{bmatrix} x \\ y \\ z \end{bmatrix}$$

---

*Section 4.1 has been revised. See M. Bouten: *Physica 42*, 572 (1969) and A. A. Wolf: *Am. J. Phys. 37*, 531 (1969).

or

(4.1.2) $\quad D(0\,0\,\gamma)D(0\,\beta\,0)D(\alpha\,0\,0)\mathbf{r} \equiv \mathbf{D}(\alpha\,\beta\,\gamma)\mathbf{r} \equiv \mathbf{D}(\omega)\mathbf{r}$

where the symbol ω represents the triad $\alpha\,\beta\,\gamma$.

A scalar field will be unchanged by a change of the frame of reference; if it is expressed as a function f of the coordinate $\mathbf{r}$ of a point P in the old frame it will be a function f' of the coordinate $\mathbf{r}'$ of the same P in the new frame, where for all points P and pairs $\mathbf{r}, \mathbf{r}'$ we have

(4.1.3) $\quad\quad\quad\quad\quad\quad\quad f'(\mathbf{r}') = f(\mathbf{r})$

We define rotation operators $D(\omega)$ on such functions f which correspond to the rotation operators $\mathbf{D}(\omega)$ on the coordinates: $f' = D(\omega)\,f$, so that

(4.1.4) $\quad\quad\quad\quad\quad D(\omega)\,f[\mathbf{D}(\omega)\mathbf{r}] = f(\mathbf{r})$

Consider the composition of two successive rotations, e.g. ω_1 followed by ω_2: $\mathbf{D}(\omega_2)\mathbf{D}(\omega_1)\mathbf{r} = \mathbf{D}(\omega)\mathbf{r}$. Then, putting $f' = D(\omega_1)f$, $\mathbf{r}' = \mathbf{D}(\omega_1)\mathbf{r}$,

$$D(\omega_2)\,f'[\mathbf{D}(\omega_2)\mathbf{r}'] = f'(\mathbf{r}') = f(\mathbf{r}) = \\ D(\omega_2)D(\omega_1)\,f[\mathbf{D}(\omega_2)\mathbf{D}(\omega_1)\mathbf{r}] = D(\omega)\,f[\mathbf{D}(\omega)\mathbf{r}]$$

I.e. $D(\omega_2)\,D(\omega_1) = D(\omega)$ and there is an isomorphism between the two sets of operators D and $\mathbf{D}$. Thus the successive application of the Euler rotations 1), 2) and 3) (see p. 7 and (4.1.2)) will correspond to transformations of the functions f given by

(4.1.5) $\quad\quad D(0\,0\,\gamma)\,D(0\,\beta\,0)\,D(\alpha\,0\,0)\,f = D(\omega)\,f.$

Now we refer to (2.1.4) and (4.1.4) and get

$$\frac{i}{\hbar}L_z f(\mathbf{r}) = \frac{\partial}{\partial\varphi}f(\mathbf{r}) = \left.\frac{\partial D(\alpha\,0\,0)}{\partial\alpha}\right|_{\alpha=0} f(\mathbf{r}).$$

Similarly

$$\frac{i}{\hbar}L_y f(\mathbf{r}) = \left.\frac{\partial D(0\,\beta\,0)}{\partial\beta}\right|_{\beta=0} f(\mathbf{r}).$$

We can now express the rotations in terms of unitary operators on the function space; integration with respect to α and β gives us

$$D(\alpha\, 0\, 0)\, f(\mathbf{r}) = \exp\frac{i\alpha}{\hbar} L_z \cdot f(\mathbf{r})$$

$$D(0\, \beta\, 0)\, f(\mathbf{r}) = \exp\frac{i\beta}{\hbar} L_y \cdot f(\mathbf{r}).$$

We get also

$$D(0\, 0\, \gamma)\, f(\mathbf{r}) = \exp\frac{i\gamma}{\hbar} L_z \cdot f(\mathbf{r}).$$

Equation (4.1.5) then gives

(4.1.8) $$D(\alpha\, \beta\, \gamma) = \exp\frac{i\gamma}{\hbar} L_z\, \exp\frac{i\beta}{\hbar} L_y\, \exp\frac{i\alpha}{\hbar} L_z$$

Now the properties of the $D(\alpha\, \beta\, \gamma)$ are determined by the algebraic properties of the operators L_x, L_y, L_z; i.e. by their commutation relations. These relations are the same for the more general operators J_x, J_y, J_z. We therefore get a representation of the finite rotations when we replace the L's by J's. That is, we may write

(4.1.9) $$D(\alpha\, \beta\, \gamma) = \exp\frac{i\gamma}{\hbar} J_z\, \exp\frac{i\beta}{\hbar} J_y\, \exp\frac{i\alpha}{\hbar} J_z$$

THE MATRIX ELEMENTS OF FINITE ROTATIONS. It is convenient to write the matrix elements of $D(\alpha\beta\gamma)$ in a more compact form; we put

(4.1.10) $$(j\, m'|D(\alpha\, \beta\, \gamma)|j\, m) \equiv \mathfrak{D}^{(j)}_{m'm}(\alpha\, \beta\, \gamma)$$

and frequently represent the three Euler angles by one symbol ω. The matrix of $D(\alpha\, \beta\, \gamma)$ in the representation $\mathfrak{D}^{(j)}$ may be symbolized by $\mathfrak{D}^{(j)}(\alpha\, \beta\, \gamma)$. We shall also write

(4.1.11) $$\mathfrak{D}^{(j)}_{m'm}(0\, \beta\, 0) \equiv d^{(j)}_{m'm}(\beta)$$

We deal with representations in which the matrices of J_z are diagonal; therefore

(4.1.12) $$\mathfrak{D}^{(j)}_{m'm}(\alpha\, \beta\, \gamma) = \exp im'\gamma\; d^{(j)}_{m'm}(\beta)\, \exp im\alpha,$$

and thus we need only consider the problem of evaluating the quantities

$$d^{(j)}_{m'm}(\beta) = \left(j\, m'\left|\exp\left(\frac{i\beta}{\hbar} J_y\right)\right|j\, m\right)$$

Let us first take the spin representation $j = \frac{1}{2}$. (2.3.19) shows that the 2×2 unitary matrix $\exp(i\beta/\hbar)J_y$ is equal to

$$\exp M \quad \text{where} \quad M = \frac{\beta}{2}\begin{pmatrix} 0 & 1 \\ -1 & 0 \end{pmatrix}$$

It is easy to demonstrate that

$$M^{2n} = (-1)^n \left(\frac{\beta}{2}\right)^{2n} \begin{pmatrix} 1 & 0 \\ 0 & 1 \end{pmatrix}$$

and

$$M^{2n+1} = (-1)^n \left(\frac{\beta}{2}\right)^{2n+1} \begin{pmatrix} 0 & 1 \\ -1 & 0 \end{pmatrix}$$

where n is an integer. Since

$$\exp M = 1 + \frac{M}{1!} + \frac{M^2}{2!} + \frac{M^3}{3!} + \cdots$$

we have

(4.1.13)
$$\exp\left(\frac{i\beta}{\hbar} J_y\right) = \begin{array}{c|cc} {}_{m'}\diagdown{}^{m} & +\tfrac{1}{2} & -\tfrac{1}{2} \\ \hline +\tfrac{1}{2} & \cos \beta/2 & \sin \beta/2 \\ -\tfrac{1}{2} & -\sin \beta/2 & \cos \beta/2 \end{array}$$

This gives, with (4.1.12), the transformation property of the spinors under rotations of the frame of coordinates. This result may be obtained in another way, namely by defining the spinors in terms of the stereographic projection of the surface of a sphere onto a plane. The points of the plane are described by homogeneous complex coordinates. Rotations of the sphere induce transformations on these coordinates, which transform in the same way as the spinors.[1]

The spin matrices σ_x, σ_y, σ_z (2.3.20), being proportional to the angular momentum operators in the $\mathfrak{D}^{(\frac{1}{2})}$ representation, may be expected to transform under rotations like the components of a vector. The reader may show for himself by working out

$$\mathfrak{D}^{(\frac{1}{2})}(\omega)\sigma_x\mathfrak{D}^{(\frac{1}{2})-1}(\omega), \text{ etc.}$$

that this is indeed the case.

We may now construct a generating function for the $d^{(j)}_{m'm}(\beta)$ for general j by making use of the representation of the $u(jm)$ in terms of the spinors (2.6.6).

We have

$$D(\alpha\,\beta\,\gamma)u(j\ m) = \frac{(\chi'_+)^{j+m}(\chi'_-)^{j-m}}{[(j+m)!(j-m)!]^{\frac{1}{2}}} = \sum_{m'} u(j\ m')\mathfrak{D}^{(j)}_{m'm}(\alpha\,\beta\,\gamma)$$

---

[1] Cf. Whittaker (1917), Weyl (1931) Chap. III §8.

For the special case $D(0\,\beta\,0)$ we have, from (4.1.13)

$$D(0\,\beta\,0)u(j\,m)$$

(4.1.14)
$$= \frac{\left(\chi_+ \cos\frac{\beta}{2} - \chi_- \sin\frac{\beta}{2}\right)^{j+m}\left(\chi_+ \sin\frac{\beta}{2} + \chi_- \cos\frac{\beta}{2}\right)^{j-m}}{[(j+m)!(j-m)!]^{\frac{1}{2}}}$$

$$= \sum_{m'} \frac{\chi_+^{j+m'}\chi_-^{j-m'}}{[(j+m')!(j-m')!]^{\frac{1}{2}}}\, d^{(j)}_{m'm}(\beta)$$

That is

(4.1.15)
$$d^{(j)}_{m'm}(\beta) = \left[\frac{(j+m')!(j-m')!}{(j+m)!(j-m)!}\right]^{\frac{1}{2}}$$
$$\cdot \sum_\sigma \binom{j+m}{j-m'-\sigma}\binom{j-m}{\sigma}(-1)^{j-m'-\sigma}$$
$$\cdot \left(\cos\frac{\beta}{2}\right)^{2\sigma+m'+m}\left(\sin\frac{\beta}{2}\right)^{2j-2\sigma-m'-m}$$

Equation (4.1.15) gives us, for example, in the case $j = 1$, the matrix $d^{(1)}(\beta)$:

m' \ m	$+1$	0	-1
$+1$	$\frac{1}{2}(1 + \cos\beta)$	$\frac{1}{\sqrt{2}}\sin\beta$	$\frac{1}{2}(1 - \cos\beta)$
0	$-\frac{1}{\sqrt{2}}\sin\beta$	$\cos\beta$	$\frac{1}{\sqrt{2}}\sin\beta$
-1	$\frac{1}{2}(1 - \cos\beta)$	$-\frac{1}{\sqrt{2}}\sin\beta$	$\frac{1}{2}(1 + \cos\beta)$

This function may be expressed in terms of the Jacobi polynomial, the properties of which will now be described briefly.

THE JACOBI POLYNOMIAL. The notation of Szegö[2] $P_n^{(\alpha,\beta)}(x)$ is used for the normalized orthogonal polynomials defined by the scalar product

(4.1.16)
$$(\varphi_n, \varphi_m) \equiv \int_{-1}^{1} (1-x)^\alpha (1+x)^\beta \varphi_n(x)\varphi_m(x)\,dx$$

The real quantities α and β should each exceed -1 if this expression is to be integrable.

---

[2] Szegö (1939), Erdelyi (1953).

The $P_n^{(\alpha,\beta)}(x)$ are normalized so that

(4.1.17)
$$P_n^{(\alpha,\beta)}(1) = \binom{n+\alpha}{n}$$

They satisfy the Rodrigues formula

(4.1.18)
$$P_n^{(\alpha,\beta)}(x) = \frac{(-1)^n}{2^n n!}(1-x)^{-\alpha}(1+x)^{-\beta} \cdot \frac{d^n}{dx^n}[(1-x)^{\alpha+n}(1+x)^{\beta+n}]$$

from which we may obtain the series expression

(4.1.19) $$P_n^{(\alpha,\beta)}(x) = 2^{-n} \sum_{\nu=0}^{n} \binom{n+\alpha}{\nu}\binom{n+\beta}{n-\nu}(x-1)^{n-\nu}(x+1)^{\nu}$$

They have the symmetry relation

(4.1.20) $$P_n^{(\alpha,\beta)}(-x) = (-1)^n P_n^{(\beta,\alpha)}(x)$$

and satisfy the differential equation

(4.1.21)
$$(1-x^2)\frac{d^2 y}{dx^2} + [\beta - \alpha - (\alpha + \beta + 2)x]\frac{dy}{dx}$$
$$+ n(n+\alpha+\beta+1)y = 0$$

The scalar product, with the above normalization, has the value

(4.1.22)
$$\int_{-1}^{1}(1-x)^{\alpha}(1+x)^{\beta}P_n^{(\alpha,\beta)}(x)P_m^{(\alpha,\beta)}(x)\,dx$$
$$= \frac{2^{\alpha+\beta+1}}{2n+\alpha+\beta+1}\frac{\Gamma(n+\alpha+1)\Gamma(n+\beta+1)}{\Gamma(n+1)\Gamma(n+\alpha+\beta+1)}\cdot\delta_{nm}$$

a result obtained by partial integration using the Rodrigues' formula.

RELATIONS BETWEEN THE MATRIX ELEMENTS OF FINITE ROTATIONS AND THE JACOBI POLYNOMIALS. Comparison of (4.1.15) and (4.1.19) gives the matrix element in terms of the Jacobi polynomial:

(4.1.23)
$$d_{m'm}^{(j)}(\beta) = \left[\frac{(j+m')!(j-m')!}{(j+m)!(j-m)!}\right]^{\frac{1}{2}}$$
$$\cdot \left(\cos\frac{\beta}{2}\right)^{m'+m}\left(\sin\frac{\beta}{2}\right)^{m'-m} P_{j-m'}^{(m'-m,m'+m)}(\cos\beta).$$

This relation is strictly speaking only valid for non-negative values of $m'-m$ and $m'+m$. Nevertheless all the results to be derived from it are true for the general case, as may be checked by making use of the symmetry properties of the $d_{m'm}^{(j)}(\beta)$ which will be discussed shortly.

The value of $d^{(l)}_{m0}(\beta)$ may be obtained easily by consideration of the above expression. We make use of the Rodrigues formulas for the Jacobi polynomial (4.1.18) and for the associated Legendre function (2.5.17) to show that

$$P^{(m,m)}_{l-m}(x) = (-2)^m \frac{l!}{(l-m)!}(1-x^2)^{-m/2} P^{-m}_l(x)$$

Hence

(4.1.24)
$$d^{(l)}_{m0}(\beta) = (-1)^m \left[\frac{(l+m)!}{(l-m)!}\right]^{\frac{1}{2}} P^{-m}_l(\cos\beta)$$
$$= \left[\frac{(l-m)!}{(l+m)!}\right]^{\frac{1}{2}} P^m_l(\cos\beta)$$

(see (2.5.18))

It follows from (4.1.12) and (2.5.29) that

(4.1.25)
$$\mathfrak{D}^{(l)}_{m0}(\alpha\beta\gamma) = (-1)^m \left(\frac{4\pi}{2l+1}\right)^{\frac{1}{2}} Y_{lm}(\beta\alpha);$$
$$\mathfrak{D}^{(l)}_{0m}(\alpha\beta\gamma) = \left(\frac{4\pi}{2l+1}\right)^{\frac{1}{2}} Y_{lm}(\beta\gamma)$$

(the second relation is obtained by use of the symmetry property (4.2.6)). In particular

(4.1.26) $$\mathfrak{D}^{(l)}_{00}(\alpha\beta\gamma) = P_l(\cos\beta)$$

We obtain from (4.1.15) another simple expression for a special choice of arguments:

(4.1.27) $$d^{(j)}_{mj}(\beta) = (-1)^{j-m} \left[\frac{(2j)!}{(j+m)!(j-m)!}\right]^{\frac{1}{2}} \left(\cos\frac{\beta}{2}\right)^{j+m} \left(\sin\frac{\beta}{2}\right)^{j-m}$$

4.2. The Symmetries of the $\mathfrak{D}^{(j)}_{m'm}$

These symmetries are found by use of the matrix elements of the rotation $(0\pi 0)$. Equation (4.1.14) shows immediately that

(4.2.1) $$d^{(j)}_{m'm}(\pi) = (-1)^{j+m} \delta_{m', -m}; \qquad d^{(j)}_{m'm}(-\pi) = (-1)^{j-m} \delta_{m', -m}.$$

We note that for the half-odd-integer representations, $D(\pi)D(\pi)$ is not equal to $D(0)$; this is an illustration of the fact that in spin representations there is a two-to-one, rather than one-to-one, correspondence between the matrices and the rotations they represent, i.e. the overall sign of a matrix in a spin representation may be reversed and the resulting matrix still represents the same rotation. This ambiguity can have no physical significance. The $d^{(j)}_{m'm}$ are the elements of unitary matrices and, moreover, are real; therefore

(4.2.2) $$d^{(j)}_{m'm}(-\beta) = d^{(j)}_{mm'}(\beta)$$

Now the successive application of rotations to a frame of coordinates corresponds to multiplication of the appropriate matrices which represent the rotations; we have therefore

$$d^{(j)}_{m'm}(\pi + \beta) = \sum_{m''} d^{(j)}_{m'm''}(\pi) d^{(j)}_{m''m}(\beta) = (-1)^{j-m'} d^{(j)}_{-m'm}(\beta) \tag{4.2.3}$$

Similarly

$$d^{(j)}_{m'm}(\pi - \beta) = (-1)^{j-m'} d^{(j)}_{-m'm}(-\beta) = (-1)^{j-m'} d^{(j)}_{m-m'}(\beta) \tag{4.2.4}$$

Now we have

$$d^{(j)}_{m'm}(\beta) = \sum_{m''} d^{(j)}_{m'm''}(\beta + \pi) d^{(j)}_{m''m}(-\pi) = (-1)^{j-m} d^{(j)}_{m'-m}(\beta + \pi)$$

Hence

$$d^{(j)}_{m'm}(\beta) = (-1)^{m'-m} d^{(j)}_{-m'-m}(\beta) \tag{4.2.5}$$

Similarly

$$d^{(j)}_{m'm}(\beta) = (-1)^{m'-m} d^{(j)}_{mm'}(\beta) \tag{4.2.6}$$

It is a simple matter to extend the symmetry relations to include the complete matrix elements $\mathfrak{D}^{(j)}_{m'm}(\alpha\,\beta\,\gamma)$ by use of (4.1.12). For example, we use (4.2.4) to get

$$\mathfrak{D}^{(j)}_{m'm}(\alpha\,\beta\,\gamma) = (-1)^{j+m'} \mathfrak{D}^{(j)}_{-m'm}(-\alpha, \beta+\pi, \gamma)$$

In particular the complex conjugate of a matrix element is given by

$$\mathfrak{D}^{(j)*}_{m'm}(\alpha\,\beta\,\gamma) = \mathfrak{D}^{(j)}_{m'm}(-\alpha\,\beta\,-\gamma) = (-1)^{m'-m} \mathfrak{D}^{(j)}_{-m'-m}(\alpha\,\beta\,\gamma) \tag{4.2.7}$$

4.3. Products of the $\mathfrak{D}^{(j)}_{m'm}(\alpha\,\beta\,\gamma)$

The products dealt with here are of the type $\mathfrak{D}^{(j_1)}_{m_1'm_1}(\alpha\,\beta\,\gamma)\mathfrak{D}^{(j_2)}_{m_2'm_2}(\alpha\,\beta\,\gamma)$. Note that the values of the Euler angles in the two matrix elements are the same. It is clear that such quantities are the matrix elements of transformation of products of angular momentum eigenvectors of the type $u(j_1m_1)u(j_2m_2)$. The results of Chapter 3 may be used when we remember that the reduction of products of the $u(jm)$ by use of the vector-coupling coefficients corresponds to a similarity transformation for the corresponding matrix elements. That is,

$$\mathfrak{D}^{(j_1)}_{m_1'm_1}(\omega)\mathfrak{D}^{(j_2)}_{m_2'm_2}(\omega) \tag{4.3.1}$$
$$= \sum_j (j_1\ m_1'\ j_2\ m_2'|j_1\ j_2\ j\ m_1'+m_2')\mathfrak{D}^{(j)}_{m_1'+m_2',\ m_1+m_2}(\omega)$$
$$\times (j_1\ j_2\ j\ m_1+m_2|j_1\ m_1\ j_2\ m_2)$$

where the values of j on the right are given by the angular momentum addition rules Substitution of the 3-j symbols for the vector-coupling coefficients gives[3]

(4.3.2)
$$\mathfrak{D}^{(j_1)}_{m_1'm_1}(\omega)\mathfrak{D}^{(j_2)}_{m_2'm_2}(\omega) = \sum_{jm'm}(2j+1)\begin{pmatrix}j_1 & j_2 & j \\ m_1' & m_2' & m'\end{pmatrix}\mathfrak{D}^{(j)*}_{m'm}(\omega)\begin{pmatrix}j_1 & j_2 & j \\ m_1 & m_2 & m\end{pmatrix}$$

which leads, as a result of the unitary property of the matrices, to the symmetric expression

(4.3.3)
$$\sum_{m_1'm_2'm_3'}\mathfrak{D}^{(j_1)}_{m_1'm_1}(\omega)\mathfrak{D}^{(j_2)}_{m_2'm_2}(\omega)\mathfrak{D}^{(j_3)}_{m_3'm_3}(\omega)$$
$$\times \begin{pmatrix}j_1 & j_2 & j_3 \\ m_1' & m_2' & m_3'\end{pmatrix} = \begin{pmatrix}j_1 & j_2 & j_3 \\ m_1 & m_2 & m_3\end{pmatrix}$$

The inverse transformation to (4.3.2) is given by

(4.3.4)
$$\sum_{\substack{m_1'm_2'\\m_1m_2}}\begin{pmatrix}j_1 & j_2 & j \\ m_1' & m_2' & m'\end{pmatrix}\mathfrak{D}^{(j_1)}_{m_1'm_1}(\omega)\mathfrak{D}^{(j_2)}_{m_2'm_2}(\omega)\begin{pmatrix}j_1 & j_2 & j' \\ m_1 & m_2 & m\end{pmatrix}$$
$$= \frac{\delta_{j'j}}{2j+1}\mathfrak{D}^{(j)*}_{m'm}(\omega)$$

4.4. Recursion Relation for the $d^{(j)}_{m'm}(\beta)$

We may specialize the formulas of the previous section to the case $j_1 = j - \frac{1}{2}$, $j_2 = \frac{1}{2}$, $j = j$. The 3-j symbols may then be evaluated by use of table 2 to give

(4.4.1)
$$d^{(j)}_{m'm}(\beta) = \left(\frac{j-m'}{j-m}\right)^{\frac{1}{2}} d^{(j-\frac{1}{2})}_{m'+\frac{1}{2}m+\frac{1}{2}}(\beta) \cdot \cos\frac{\beta}{2}$$
$$- \left(\frac{j+m'}{j-m}\right)^{\frac{1}{2}} d^{(j-\frac{1}{2})}_{m'-\frac{1}{2}m+\frac{1}{2}}(\beta) \cdot \sin\frac{\beta}{2}$$

This relation is of course useless when $m = j$; in this case we use (4.1.27).

4.5. Computation[4] of the $d^{(j)}_{m'm}(\beta)$

A similarity transformation may be employed to express a $D(0\,\beta\,0)$ in terms of a rotation about the z-axis, i.e. a $D(\xi\,0\,0)$, which is diagonal in our representations:

(4.5.1) $$D(0\,\beta\,0) = D\left(-\frac{\pi}{2}\,0\,0\right)D\left(0\,-\frac{\pi}{2}\,0\right)D(\beta\,0\,0)D\left(0\,\frac{\pi}{2}\,0\right)D\left(\frac{\pi}{2}\,0\,0\right)$$

---
[3] See (4.6.5) for specialization of this result to spherical harmonics.
[4] Based on method of Wigner (1951).

Thus the problem of computing any matrix $d^{(j)}(\beta)$ is reduced to that of computing the one matrix $d^{(j)}(\pi/2)$ which we symbolize by $\Delta^{(j)}$. These matrices may be built up by use of the recursion relation (4.4.1) and a number of them are exhibited in Table 4. In the $\mathfrak{D}^{(j)}$ representation (4.5.1) gives

(4.5.2)
$$d_{m'm}^{(j)}(\beta) = \sum_{m''} e^{im'\pi/2} \Delta_{m''m'}^{(j)} e^{-im''\beta} \Delta_{m''m}^{(j)} e^{-im\pi/2}$$
$$= \Delta_{0m'}^{(j)} \Delta_{0m}^{(j)} \kappa(0) + 2 \sum_{m''>0} \Delta_{m''m'}^{(j)} \Delta_{m''m}^{(j)} \kappa(m''\beta)$$

where

(4.5.3)
$$\begin{aligned}\kappa(x) &= \cos x \text{ if } m' - m \equiv 0 \pmod 4 \\ &= \sin x \text{ if } m' - m \equiv 1 \pmod 4 \\ &= -\cos x \text{ if } m' - m \equiv 2 \pmod 4 \\ &= -\sin x \text{ if } m' - m \equiv 3 \pmod 4\end{aligned}$$

4.6. Integrals Involving the $\mathfrak{D}_{m'm}^{(j)}(\alpha\,\beta\,\gamma)$

The orthogonality and normalization of the $\mathfrak{D}_{m'm}^{(j)}(\alpha\,\beta\,\gamma)$ with respect to integrations over the Euler angles are easily checked by reference to the corresponding properties of the Jacobi polynomials.

Equation (4.1.23) shows that, putting $\cos\beta = t$,

$$\frac{1}{8\pi^2}\int_0^{2\pi}\int_0^{\pi}\int_0^{2\pi} \mathfrak{D}_{m_1'm_1}^{(j_1)*}(\alpha\,\beta\,\gamma)\mathfrak{D}_{m_2'm_2}^{(j_2)}(\alpha\,\beta\,\gamma)\,d\alpha\,\sin\beta\,d\beta\,d\gamma$$
$$= \frac{\delta_{m_1'm_2'}\delta_{m_1m_2}}{2}\left[\frac{(j_1+m_1')!(j_1-m_1')!(j_2+m_1')!(j_2-m_1')!}{(j_1+m_1)!(j_1-m_1)!(j_2+m_1)!(j_2-m_1)!}\right]^{\frac{1}{2}}$$
$$\cdot \int_{-1}^{1}\left(\frac{1+t}{2}\right)^{m_1'+m_1}\left(\frac{1-t}{2}\right)^{m_1'-m_1} P_{j_1-m_1'}^{(m_1'-m_1,m_1'+m_1)}(t) P_{j_2-m_1'}^{(m_1'-m_1,m_1'+m_1)}(t)\,dt$$

which is by (4.1.22) equal to $\delta_{m_1'm_2'}\delta_{m_1m_2}\delta_{j_1j_2}\cdot 1/(2j_1+1)$. That is,

(4.6.1)
$$\frac{1}{8\pi^2}\int_0^{2\pi}\int_0^{\pi}\int_0^{2\pi} \mathfrak{D}_{m_1'm_1}^{(j_1)*}(\alpha\,\beta\,\gamma)\mathfrak{D}_{m_2'm_2}^{(j_2)}(\alpha\,\beta\,\gamma)\,d\alpha\,\sin\beta\,d\beta\,d\gamma$$
$$= \delta_{m_1'm_2'}\delta_{m_1m_2}\delta_{j_1j_2}\cdot\frac{1}{2j_1+1}$$

Application of this result to (4.3.2) gives the symmetric expression for the integral over the product of three $\mathfrak{D}$'s:

(4.6.2)
$$\frac{1}{8\pi^2}\int_0^{2\pi}\int_0^{\pi}\int_0^{2\pi} \mathfrak{D}_{m_1'm_1}^{(j_1)}(\alpha\,\beta\,\gamma)\mathfrak{D}_{m_2'm_2}^{(j_2)}(\alpha\,\beta\,\gamma)$$
$$\times \mathfrak{D}_{m_3'm_3}^{(j_3)}(\alpha\,\beta\,\gamma)\,d\alpha\,\sin\beta\,d\beta\,d\gamma$$
$$= \begin{pmatrix} j_1 & j_2 & j_3 \\ m_1' & m_2' & m_3' \end{pmatrix}\begin{pmatrix} j_1 & j_2 & j_3 \\ m_1 & m_2 & m_3 \end{pmatrix}$$

4.6 · INTEGRALS INVOLVING THE $\mathfrak{D}^{(j)}_{m'm}(\alpha\beta\gamma)$

We may use (4.1.25) to specialize this integral to one over three spherical harmonics:

$$
(4.6.3) \quad \int_0^{2\pi} \int_0^{\pi} Y_{l_1 m_1}(\theta, \varphi) Y_{l_2 m_2}(\theta, \varphi) Y_{l_3 m_3}(\theta, \varphi) \sin\theta \, d\theta \, d\varphi
$$
$$
= \left[\frac{(2l_1+1)(2l_2+1)(2l_3+1)}{4\pi}\right]^{\frac{1}{2}} \begin{pmatrix} l_1 & l_2 & l_3 \\ 0 & 0 & 0 \end{pmatrix} \begin{pmatrix} l_1 & l_2 & l_3 \\ m_1 & m_2 & m_3 \end{pmatrix}
$$

Further specialization gives the integral over three Legendre functions:

$$
(4.6.4) \quad \frac{1}{2}\int_0^{\pi} P_{l_1}(\cos\theta) P_{l_2}(\cos\theta) P_{l_3}(\cos\theta) \sin\theta \, d\theta = \begin{pmatrix} l_1 & l_2 & l_3 \\ 0 & 0 & 0 \end{pmatrix}^2
$$

The 3-j symbols with $m_1 = m_2 = m_3 = 0$ may be evaluated by the methods discussed in (3.8).

Equation (4.3.2) may be specialized in the same way to give an expression for the product of two spherical harmonics which have the same angles for arguments.

$$
(4.6.5) \quad Y_{l_1 m_1}(\theta,\varphi) Y_{l_2 m_2}(\theta,\varphi) = \sum_{lm} \left[\frac{(2l_1+1)(2l_2+1)(2l+1)}{4\pi}\right]^{\frac{1}{2}}
$$
$$
\times \begin{pmatrix} l_1 & l_2 & l \\ m_1 & m_2 & m \end{pmatrix} Y^*_{lm}(\theta,\varphi) \begin{pmatrix} l_1 & l_2 & l \\ 0 & 0 & 0 \end{pmatrix}
$$

Suppose the rotation $(\alpha\,\beta\,\gamma)$ is the result of the successive application of, in that order, $(\alpha_1\,\beta_1\,\gamma_1)$ and $(\alpha_2\,\beta_2\,\gamma_2)$. Then we have

$$
\mathfrak{D}^{(j)}_{m'm}(\alpha\,\beta\,\gamma) = \sum_{m''} \mathfrak{D}^{(j)}_{m'm''}(\alpha_2\,\beta_2\,\gamma_2) \mathfrak{D}^{(j)}_{m''m}(\alpha_1\,\beta_1\,\gamma_1)
$$

The spherical harmonic addition theorem[5] is obtained by setting $j = l =$ integer and $m' = m = 0$. Then

$$
(4.6.6) \quad P_l(\cos\omega) = \frac{4\pi}{2l+1} \sum_m Y^*_{lm}(\theta,\varphi) Y_{lm}(\theta',\varphi')
$$

where $\cos\omega = \cos\theta\cos\theta' + \sin\theta\sin\theta'\cos(\varphi-\varphi')$. In terms of the function $C^{(l)}_m$ (cf. (2.5.31)) we have

$$
(4.6.7) \quad P_l(\cos\omega) = \sum_m C^{(l)*}_m(\theta,\varphi) C^{(l)}_m(\theta',\varphi')
$$
$$
= \sum_m (-1)^m C^{(l)}_m(\theta,\varphi) C^{(l)}_{-m}(\theta',\varphi')
$$

---

[5] Cf. Whittaker and Watson (1946) p. 328, Condon and Shortley (1935) p. 53.

4.7. The $\mathfrak{D}^{(j)}_{m'm}(\omega)$ as Angular Momentum Eigenfunctions

Let us return to the consideration of the differential properties of finite rotation operators with respect to variation of the parameters $\alpha\beta\gamma$ (cf. (4.1)). We have seen that

(4.7.1) $$\frac{\partial}{\partial\alpha} D(\alpha\,\beta\,\gamma) = D(\alpha\,\beta\,\gamma)\frac{\partial}{\partial\varphi}$$

In the same way

(4.7.2) $$\frac{\partial}{\partial\beta} D(\alpha\,\beta\,\gamma) = D(\alpha\,\beta\,\gamma)\frac{\partial}{\partial\varphi_\beta}$$

where φ_β is the angle coordinate of a point in the frame of reference measured about the line of nodes (the y axis in S'), and

(4.7.3) $$\frac{\partial}{\partial\gamma} D(\alpha\,\beta\,\gamma) = D(\alpha\,\beta\,\gamma)\frac{\partial}{\partial\varphi_\gamma}$$

where φ_γ is the angle measured about the figure axis (the z axis in S''). Thus to each differential operation x on the $D(\alpha\,\beta\,\gamma)$ with respect to α, β, γ corresponds a differential operation ξ on a function defined in coordinate space. Clearly we may write the result of application of a succession of operations $x_1, x_2, x_3, \ldots$

(4.7.4) $$\cdots x_3 x_2 x_1 D(\alpha\,\beta\,\gamma) = D(\alpha\,\beta\,\gamma) \cdots \xi_3\xi_2\xi_1$$

Now in (4.7.1), (4.7.2), and (4.7.3) we have made use of the angles φ, φ_β, and φ_γ to emphasize that the differential operators on the right-hand side work upon functions defined in coordinate space. They are really identical with α, β, and γ respectively, and so we see from these equations and from (4.7.4) that the representation of infinitesimal rotations given by the $\mathfrak{D}$'s corresponds to that given by the angular momentum eigenvectors considered in (2.2).

We may make use of (2.2.2) and write

$$L_x D(\alpha\,\beta\,\gamma) = -i\hbar\left\{-\cos\alpha\cot\beta\frac{\partial}{\partial\alpha} - \sin\alpha\frac{\partial}{\partial\beta} + \frac{\cos\alpha}{\sin\beta}\frac{\partial}{\partial\gamma}\right\}D(\alpha\,\beta\,\gamma)$$

$$= D(\alpha\,\beta\,\gamma)L_x\;;$$

$$L_y D(\alpha\,\beta\,\gamma) = -i\hbar\left\{-\sin\alpha\cot\beta\frac{\partial}{\partial\alpha} + \cos\alpha\frac{\partial}{\partial\beta} + \frac{\sin\alpha}{\sin\beta}\frac{\partial}{\partial\gamma}\right\}D(\alpha\,\beta\,\gamma)$$

$$= D(\alpha\,\beta\,\gamma)L_y\;;$$

$$L_z D(\alpha\,\beta\,\gamma) = -i\hbar\frac{\partial}{\partial\alpha} D(\alpha\,\beta\,\gamma) = D(\alpha\,\beta\,\gamma)L_z$$

It follows that

$$\mathbf{L}^2 D(\alpha\,\beta\,\gamma) = \hbar^2 \left\{ -\frac{\partial^2}{\partial\beta^2} - \cot\beta\,\frac{\partial}{\partial\beta} \right.$$
$$\left. - \frac{1}{\sin^2\beta}\left(\frac{\partial^2}{\partial\alpha^2} + \frac{\partial^2}{\partial\gamma^2} - 2\cos\beta\,\frac{\partial^2}{\partial\alpha\,\partial\gamma}\right) \right\} D(\alpha\,\beta\,\gamma)$$
$$= D(\alpha\,\beta\,\gamma)\mathbf{L}^2$$

When we remember that the eigenvalue of $\mathbf{L}^2$ is, according to (2.3.15) $\hbar^2 l(l+1)$, we see that we have constructed an eigenvalue equation for $\mathbf{L}^2$; the matrix elements of $D(\alpha\,\beta\,\gamma)$ are clearly the eigenfunctions of $\mathbf{L}^2$ and L_z. That is, we may rewrite the above equation in the form

(4.7.5)
$$\left\{\frac{\partial^2}{\partial\beta^2} + \cot\beta\,\frac{\partial}{\partial\beta} + \frac{1}{\sin^2\beta}\left(\frac{\partial^2}{\partial\alpha^2} + \frac{\partial^2}{\partial\gamma^2} - 2\cos\beta\,\frac{\partial^2}{\partial\alpha\,\partial\gamma}\right) + l(l+1)\right\} \mathfrak{D}^{(l)}_{mk}(\alpha\,\beta\,\gamma) = 0$$

where we have chosen the $l\,m$; $l\,k$ matrix component of the operator equation. Thus $\mathfrak{D}^{(l)}_{mk}(\alpha\,\beta\,\gamma)$ is the eigenfunction of $\mathbf{L}^2$ with eigenvalue $\hbar^2 l(l+1)$ and the eigenfunction of L_z with eigenvalue $\hbar m$. It is simultaneously an eigenfunction with eigenvalue $\hbar k$ of the angular momentum operator $-i\hbar(\partial/\partial\gamma)$. This operator is the analogue of L_z in the moving coordinate system and commutes with $\mathbf{L}^2$ and L_z.

It has been remarked by Bopp and Haag (1950) that the $\mathfrak{D}^{(j)}_{mk}(\alpha\,\beta\,\gamma)$ with half-odd-integer j may be regarded as eigenfunctions of $\mathbf{L}^2$ and L_z, although these are defined by (2.2.3) and (2.2.2) in terms of differential operators; a more concrete representation of the spin eigenvectors is thus obtained.

(4.7.5) gives a differential equation for $d^{(l)}_{mk}(\beta)$ which is defined by (4.1.12) in terms of $\mathfrak{D}^{(l)}_{mk}(\alpha\,\beta\,\gamma)$:

(4.7.6) $$\left\{\frac{d^2}{d\beta^2} + \cot\beta\,\frac{d}{d\beta} - \frac{m^2 + k^2 - 2mk\cos\beta}{\sin^2\beta} + l(l+1)\right\} d^{(l)}_{mk}(\beta) = 0$$

We pass now to the quantum mechanics of the symmetric top[6], a topic of great importance in the theories of molecular spectra[7] and of the collective model of the atomic nucleus.[8]

4.8. The Symmetric Top

The kinetic energy T of a rigid body with symmetry about the figure axis which rotates about its center of mass is given by

(4.8.1) $$2T = I_1(\omega_1^2 + \omega_2^2) + I_3\omega_3^2,$$

---
[6] Kronig and Rabi (1927), Dennison (1931), Casimir (1931).
[7] Herzberg (1939).
[8] Bohr and Mottelson (1953), (1955), Bohr (1952).

where ω_1, ω_2, and ω_3 are the angular velocities about the x, y, and z axes of the *moving* frame of reference fixed in the body, and $I_1 = I_2$ and I_3 are the corresponding moments of inertia.

The angular velocities ω_1, ω_2, and ω_3 are given in terms of the rates of change of the Euler angles by the Euler geometrical equations[9]

(4.8.2)
$$\omega_1 = \dot{\beta} \sin \gamma - \dot{\alpha} \sin \beta \cos \gamma$$
$$\omega_2 = \dot{\beta} \cos \gamma + \dot{\alpha} \sin \beta \sin \gamma$$
$$\omega_3 = \dot{\alpha} \cos \beta + \dot{\gamma}$$

(Note the difference between these equations and those of (2.2); in this case we refer to the moving axes, in the other to the fixed axes.)

Hence the kinetic energy is given in terms of the Euler angles by

(4.8.3) $$2T = I_1(\dot{\beta}^2 + \dot{\alpha}^2 \sin^2 \beta) + I_3(\dot{\alpha} \cos \beta + \dot{\gamma})^2$$

On replacement of the time derivatives of the Euler angles by the generalized momenta $p_\alpha = \partial T / \partial \dot{\alpha}$ etc., we have

(4.8.4) $$2T = \frac{p_\beta^2}{I_1} + \left(\frac{\cos^2 \beta}{I_1 \sin^2 \beta} + \frac{1}{I_3}\right) p_\gamma^2 + \frac{1}{I_1 \sin^2 \beta} p_\alpha^2 - \frac{2 \cos \beta}{I_1 \sin^2 \beta} p_\alpha p_\gamma$$

The Schrödinger equation for the system is obtained by the substitutions $p_\alpha \to -i\hbar(\partial/\partial\alpha)$ etc.:

(4.8.5)
$$\frac{-\hbar^2}{2I_1} \left\{ \frac{\partial^2}{\partial \beta^2} + \cot \beta \frac{\partial}{\partial \beta} + \left(\frac{I_1}{I_3} + \cot^2 \beta\right) \frac{\partial^2}{\partial \gamma^2} \right.$$
$$\left. + \frac{1}{\sin^2 \beta} \cdot \frac{\partial^2}{\partial \alpha^2} - \frac{2 \cos \beta}{\sin^2 \beta} \cdot \frac{\partial^2}{\partial \alpha \partial \gamma} \right\} \Psi(\alpha \beta \gamma) = E \Psi(\alpha \beta \gamma).$$

If we set $I_1 = I_3$, i.e. consider a rigid body whose ellipsoid of inertia has spherical symmetry, (4.8.5) reduces to (4.7.5). The $\mathfrak{D}_{mk}^{(l)}$ are thus clearly eigenfunctions of the corresponding Schrödinger equation. We may effect a separation of (4.8.5) even when $I_1 \neq I_3$ by the *ansatz*

(4.8.6) $$\Psi(\alpha \beta \gamma) = B(\beta) \exp i(m\alpha + k\gamma)$$

The differential equation for $B(\beta)$ is then

(4.8.7)
$$-\frac{\hbar^2}{2I_1} \left\{ \frac{d^2}{d\beta^2} + \cot \beta \frac{d}{d\beta} - \frac{m^2 + k^2 - 2mk \cos \beta}{\sin^2 \beta} \right\} B(\beta)$$
$$= \left\{ E - \frac{\hbar^2 k^2 (I_1 - I_3)}{2 I_1 I_3} \right\} B(\beta).$$

---

[9] Cf. Synge and Griffith (1949) pp. 289, 424 et seq. See also Reiche and Rademacher (1926).

We see from (4.7.6) that the $d^{(l)}_{mk}(\beta)$ are still eigenfunctions of this equation; however the energy corresponding to a given $\mathfrak{D}^{(l)}_{mk}$ is different from that in the case of the spherical rotator.

The eigenfunctions of the *asymmetric* rigid rotator are more complicated than those just considered; they may nevertheless be expressed as linear combinations of symmetric top eigenfunctions.[10]

---

[10] Mulliken (1941), King, Hainer, and Cross (1943), Van Winter (1954).

CHAPTER 5

Spherical Tensors and Tensor Operators

5.1. Spherical Tensors[1]

REDUCTION OF CARTESIAN TENSORS. We shall examine the properties of Cartesian tensors in three dimensions when they are subjected not to the whole group of linear nonsingular transformations but to the subgroup of *orthogonal* transformations. In this case a tensor of given rank which is irreducible under the full group may be reduced; it will be sufficient to illustrate this point by the simple example of a tensor of rank 2, built up by taking all 9 products of the components of two vectors **x** and **y**. A typical tensor component is thus

$$T_{ik} = x_i y_k \qquad (i, k = 1, 2, 3)$$

The tensor T_{ik} may, as is well known, be split into symmetric and antisymmetric tensors

$$S_{ik} = \tfrac{1}{2}(T_{ik} + T_{ki}); \qquad A_{ik} = \tfrac{1}{2}(T_{ik} - T_{ki})$$

Now under orthogonal transformations the scalar product $(\mathbf{x}\cdot\mathbf{y}) \equiv \sum_i x_i y_i$ is invariant; it follows that the symmetric tensor is reducible; we extract the invariant quantity and obtain

$$S'_{ik} = \tfrac{1}{2}(x_i y_k + x_k y_i - \tfrac{2}{3}(\mathbf{x}\cdot\mathbf{y})\delta_{ik})$$

A similar process may be carried out with tensors of higher rank; it amounts simply to subtracting all the quantities which are invariant under orthogonal transformations. In this way we may in principle build up irreducible tensors of any rank from the components of the basic vectors. It may also be shown that, if these tensors are constructed from the components of a single vector **r**, then they are identical, apart from constant factors, with the normalized harmonic polymonials $\mathcal{Y}_{lm}(\mathbf{r})$.

THE HARMONIC POLYNOMIALS. A harmonic polynomial[2] $H_l(\mathbf{r})$ is a homogeneous polynomial of degree l in the components x, y, and z of **r**, and which satisfies Laplace's equation

$$\Delta H_l(\mathbf{r}) = 0$$

---

[1] Cf. Rose (1954).
[2] Cf. Erdélyi (1953) Chap. XI.

5.1 · SPHERICAL TENSORS

The harmonic polynomials may be generated as follows: we take a vector **v** of zero amplitude (i.e. $(\mathbf{v}\cdot\mathbf{v}) = 0$)

$$\mathbf{v} = (-2t,\ 1 - t^2,\ i(1 + t^2))$$

Now we have

$$\Delta(\mathbf{r}\cdot\mathbf{a})^l = l(l-1)(\mathbf{a}\cdot\mathbf{a})(\mathbf{r}\cdot\mathbf{a})^{l-2} \quad \text{where } \mathbf{a} = \text{constant}.$$

Hence $\Delta(\mathbf{r}\cdot\mathbf{v})^l = 0$ and the coefficients of the powers of t in $(\mathbf{r}\cdot\mathbf{v})^l = [y + iz - 2xt - (y - iz)t^2]^l$ are themselves harmonic polynomials. That is, we have the generating function

$$(5.1.1) \qquad [y + iz - 2xt - (y - iz)t^2]^l = t^l \sum_{m=-l}^{l} H_{lm}(\mathbf{r}) t^m$$

Thus there are $2l + 1$ independent harmonic polynomials of a given degree l.

The functions $r^{-l}H_{lm}(\mathbf{r})$ are one-valued continuous functions on the unit sphere which are themselves solutions of the Laplace equation. Hence the $2l + 1$ functions $r^{-l}H_{lm}(\mathbf{r})$ ($m = 0, \pm 1, \ldots, \pm l$) are linear combinations of the $2l + 1$ spherical harmonics $Y_{lm}(\theta, \varphi)$. We shall define the normalized harmonic polymonial or solid harmonic as

$$(5.1.2) \qquad \mathcal{Y}_{lm}(\mathbf{r}) = r^l Y_{lm}(\theta, \varphi)$$

A number of the $\mathcal{Y}_{lm}(\mathbf{r})$ are presented as functions of x, y, and z in Table 1.

THE SPHERICAL TENSOR NOTATION. We define the spherical components[3] of a vector **r** as

$$(5.1.3) \qquad r_{\pm 1} = \mp \frac{1}{\sqrt{2}}(x \pm iy); \qquad r_0 = z$$

i.e. $\qquad x = \frac{1}{\sqrt{2}}(r_{-1} - r_{+1}); \qquad y = \frac{i}{\sqrt{2}}(r_{-1} + r_{+1})$

thus the solid harmonic $\mathcal{Y}_{1m}(\mathbf{r})$ is expressed in this notation as

$$(5.1.4) \qquad \mathcal{Y}_{1m}(\mathbf{r}) = \left(\frac{3}{4\pi}\right)^{\frac{1}{2}} r_m \qquad (m = \pm 1, 0)$$

We may construct similar expressions for the components of any other quantity which transforms like a vector under rotations. With the aid of this convention we may use the vector-coupling methods derived in Chapter 3 to construct spherical tensors of any rank from the spherical components of a given set of vector quantities. In general we have

$$(5.1.5) \qquad T(l\ m) = \sum_{m_1 m_2} T(l_1\ m_1) T(l_2\ m_2)(l_1\ m_1\ l_2\ m_2 | l_1\ l_2\ l\ m)$$

---

[3] Note that this convention differs from that of (2.3.1).

and the corresponding inverse expression for $T(l_1\ m_1)T(l_2\ m_2)$, (5.1.9), where $T(l_1\ m_1), \ldots$ are the components of any spherical tensors which transform under rotations in the same way as $\mathcal{Y}_{l_1 m_1}, \ldots$ respectively. Formula (4.6.5) may be adapted for the case of solid harmonics:

$$\mathcal{Y}_{l_1 m_1}(\mathbf{r})\, \mathcal{Y}_{l_2 m_2}(\mathbf{r})$$

(5.1.6)
$$= \sum_{lm} \left[\frac{(2l_1+1)(2l_2+1)}{4\pi(2l+1)}\right]^{\frac{1}{2}} (l_1\ m_1\ l_2\ m_2 | l_1\ l_2\ l\ m)$$

$$\times\ (l_1\ 0\ l_2\ 0 | l_1\ l_2\ l\ 0) r^{l_1+l_2-l}\mathcal{Y}_{lm}(\mathbf{r}).$$

Note that the only l values appearing on the right are those satisfying $l_1 + l_2 + l$ = even integer.[4]

We may make use of the unitary properties (3.5.4) of the V-C coefficients to show from (5.1.6) that a tensor operator $T(l\ m)$ formed according to (5.1.5) from two solid harmonics $\mathcal{Y}_{l_1 m_1}(\mathbf{r})$, $\mathcal{Y}_{l_2 m_2}(\mathbf{r})$ is related to $\mathcal{Y}_{lm}(\mathbf{r})$ by

(5.1.7) $\quad T(l\ m) = \left[\dfrac{(2l_1+1)(2l_2+1)}{4\pi(2l+1)}\right]^{\frac{1}{2}} (l_1\ 0\ l_2\ 0 | l_1\ l_2\ l\ 0) r^{l_1+l_2-l}\mathcal{Y}_{lm}(\mathbf{r}).$

As an example we consider the familiar cross-product of two vectors $\mathbf{x} \times \mathbf{y}$. The formulas for the V-C coefficients given in Table 5.2 are used. If we define

$$T(1\ m) = \sum_{m_1 m_2} x_{m_1} y_{m_2} (1\ m_1\ 1\ m_2 | 1\ 1\ 1\ m)$$

we get the result

$$T(1\ -1) = \frac{1}{\sqrt{2}} (-x_{-1}y_0 + x_0 y_{-1})$$

$$T(1\ 0) = \frac{1}{\sqrt{2}} (-x_{-1}y_{+1} + x_{+1}y_{-1})$$

$$T(1\ 1) = \frac{1}{\sqrt{2}} (-x_0 y_{+1} + x_{+1}y_0)$$

I.e.

(5.1.8) $\qquad T(1\ m) = \dfrac{i}{\sqrt{2}} (\mathbf{x} \times \mathbf{y})_m$

The inverse transformation to (5.1.5) is given by the orthogonality of the V-C coefficients; thus

(5.1.9) $\quad T(l_1\ m_1)T(l_2\ m_2) = \sum_{lm} T(l\ m)(l_1\ l_2\ l\ m | l_1\ m_1\ l_2\ m_2).$

---

[4] See (3.7.14).

The process of *polarization*[5] may be used to introduce new variables into spherical tensors without altering the transformation properties. Thus a new tensor of type l, m may be gotten from $\mathcal{Y}_{lm}(\mathbf{r})$ by the application up to l times of polarizing operators $\mathbf{a} \cdot \nabla$, $\mathbf{b} \cdot \nabla$, ... etc., where $\mathbf{a}, \mathbf{b}, \ldots$ are any vectors and

(5.1.10) $$\mathbf{a} \cdot \nabla \equiv a_{+1} \frac{\partial}{\partial r_{+1}} + a_0 \frac{\partial}{\partial r_0} + a_{-1} \frac{\partial}{\partial r_{-1}}.$$

The properties of the gradient operator ∇ will be examined more closely in (5.7).

5.2. Tensor Operators in Quantum Mechanics

DEFINITION OF THE TENSOR OPERATOR.. A finite rotation of the frame of reference of a quantum mechanical system about the origin may be considered to induce a canonical transformation[6] on the coordinates and momenta, and a corresponding unitary transformation on all operators relating to the system. To each rotation $(\alpha\,\beta\,\gamma) \equiv (\omega)$ corresponds a unitary transformation $D\,(\alpha\,\beta\,\gamma) \equiv D(\omega)$ and we may write for any operator Q:

$$Q \to Q' = D(\omega) Q D^{-1}(\omega)$$

We extend the concept of spherical tensor discussed in the previous section to that of an *irreducible tensor operator* $\mathsf{T}(k)$ which is a set of $2k + 1$ operators $T(k\,q)$ $(q = -k, -k+1, \ldots, k-1, k)$ which transform under rotations of the frame of coordinates like the components of the spherical tensor $\mathcal{Y}_{kq}$, namely as

(5.2.1) $$D(\omega) T(k\,q) D^{-1}(\omega) = \sum_{q'=-k}^{k} T(k\,q')\mathfrak{D}^{(k)}_{q'q}(\omega)$$

Since the operators of total angular momentum of the system are multiples of the infinitesimal rotation operators, we may replace the unitary transformation on the left by a commutator, giving for any component of angular momentum J_ξ

(5.2.2) $$[J_\xi, T(k\,q)] = \sum_{q'} T(k\,q')(k\,q'|J_\xi|k\,q)$$

i.e.

(5.2.3) $$[J_\pm, T(k\,q)] = T(k\,q\pm 1) \cdot \hbar[(k \mp q)(k \pm q + 1)]^{\frac{1}{2}}$$
$$[J_0, T(k\,q)] = T(k\,q) \cdot \hbar q$$

---

[5] Cf. Weyl (1939), Falkoff and Uhlenbeck (1950).
[6] Dirac (1947).

which are equivalent to the more compact relations

$$[\mathbf{J} \cdot [\mathbf{J}, T(k\ q)]] = \hbar^2 k(k+1) T(k\ q); \quad [J_0, T(k\ q)] = \hbar q T(k\ q)$$

(I.e. $\tfrac{1}{2}[J_+, [J_-, T(k\ q)]] + \tfrac{1}{2}[J_-, [J_+, T(k\ q)]] + [J_0, [J_0, T(k\ q)]]$
$$= \hbar^2 k(k+1) T(k\ q).)$$

These commutators correspond to the definition of tensor operator given by Racah (1942).

The most familiar examples of tensor operators are the position vector **r** and linear momentum **p** of a particle, which are tensor operators of rank 1, i.e. *vector* operators.[7] The angular momentum **J** of a system is clearly itself a vector operator. Other tensor operators arise when we consider the multipole moments of systems of particles; for example the electric quadrupole moment of the nucleus.

We may build up tensor operators of higher rank by exactly the same methods as discussed for *c*-number tensors in (5.1). It may be necessary to pay attention to symmetrization when dealing with operators whose components do not commute. The *parity* of an operator is an important quantity in quantum mechanics. It is clear that all components of a given tensor operator have the same parity for the parity operator commutes with rotations; and the parity of a tensor operator built up by the methods of (5.1) is given by the product of the parities of the constituent tensor operators. For example, the dipole moment of a system has odd, the quadrupole moment even parity.

The scalar product of two tensor operators of the same rank is represented conventionally[8] by

$$(5.2.4) \qquad (\mathbf{T} \cdot \mathbf{U}) = \sum_q (-1)^q T(k\ q) U(k\ -q)$$

and not by the expression gotten by use of (5.1.5) with $l = m = 0$; the two forms differ only by a constant factor. There are many physical problems where we encounter quantities which may be expressed as the scalar or tensor product of two tensor operators, which are usually related to different parts of the system. We shall see that such a formalism introduces a great simplification into the calculations, expecially when allied with the use of 6-j and 9-j symbols, which will be discussed in the next chapter.

EXAMPLES OF USE OF THE TENSOR OPERATOR NOTATION. The term representing the interaction between the atomic nucleus and the electric field of the surrounding electrons, which is responsible for the hyperfine structure, is given by

$$(5.2.5) \qquad H' = \sum_{i,p} \frac{e_i e_p}{|\mathbf{r}_i - \mathbf{r}_p|} = \sum_{i,p,l} \frac{e_i e_p}{r_i^{l+1}} r_p^l P_l\left(\frac{\mathbf{r}_p \cdot \mathbf{r}_i}{r_p r_i}\right)$$

[7] Cf. Güttinger and Pauli (1931), Wigner (1931), Condon and Shortley (1935).
[8] Cf. Racah (1942).

where e_i, $\mathbf{r}_i$ and e_p, $\mathbf{r}_p$ are the charges and position vectors of the electrons and protons respectively. We consider the quadrupole term ($l = 2$). The spherical harmonic addition theorem (4.6.7) makes it possible to separate the expression into functions of electron and proton coordinates:

$$(5.2.6) \quad H' = \sum_m (-1)^m \sum_p e_p r_p^l C_m^{(2)}(\theta_p, \varphi_p) \sum_i e_i r_i^{-l-1} C_{-m}^{(2)}(\theta_i, \varphi_i)$$

which is of the form of (5.2.4); the matrix elements of H' will be evaluated in Chapter 7.

The tensor product of tensor operators arises in the treatment of the so-called tensor interaction between nucleons. This interaction is usually written

$$(5.2.7) \quad S_{12} = J(r_{12})\left\{\frac{(\mathbf{\sigma}_1 \cdot \mathbf{r}_{12})(\mathbf{\sigma}_2 \cdot \mathbf{r}_{12})}{r_{12}^2} - \frac{1}{3}(\mathbf{\sigma}_1 \cdot \mathbf{\sigma}_2)\right\}$$

where $\mathbf{r}_{12}$ is the vector joining the nucleons 1 and 2 and $\mathbf{\sigma}_1$ and $\mathbf{\sigma}_2$ are their respective spin operators. It may also with advantage be written[9] as the scalar product

$$S_{12} = (\mathbf{S} \cdot \mathbf{L})$$

where $\mathbf{S}(2)$ is the irreducible tensor operator of rank 2 formed from $\mathbf{\sigma}_1$ and $\mathbf{\sigma}_2$ and $\mathbf{L}(2)$ is a product of the scalar $J(r_{12})$ and the irreducible tensor operator of rank 2 formed from the unit vector $\mathbf{r}_{12}/r_{12}$. Here again we see that the operators in the scalar product refer to different parts of the system. The spin tensor may be constructed by polarization:

$$S(2m) = \left(\frac{2\pi}{15}\right)^{\frac{1}{2}} (\mathbf{\sigma}_1 \cdot \nabla)(\mathbf{\sigma}_2 \cdot \nabla) \mathcal{Y}_{2m}(\mathbf{r})$$

The orbital tensor appears as

$$L(2m) = \left(\frac{8\pi}{15}\right)^{\frac{1}{2}} \frac{J(r_{12})}{r_{12}^2} \mathcal{Y}_{2m}(\mathbf{r}_{12})$$

5.3. Factorization of the Matrix Elements of Tensor Operators (Wigner-Eckart Theorem)[10]

Consider the component $T(k\,q)$ of a tensor operator acting on a state vector of a system which is a simultaneous eigenvector of the angular momentum operators $\mathbf{J}^2$, J_z of this system. We call the state vector $u(\gamma\,j\,m)$. Let us examine the effect of a finite rotation ω of the coordinate system on the quantity $T(kq)u(\gamma jm)$. We have

$$D(\omega)[T(k\,q)u(\gamma\,j\,m)] = [D(\omega)T(k\,q)D^{-1}(\omega)]D(\omega)u(\gamma\,j\,m)$$

---
[9] Cf. Elliott (1953).
[10] Wigner (1931), Eckart (1930).

Reference to the definition (5.2.1) of the tensor operator shows us that this is equal to

$$\sum_{q'm'} T(k\ q')u(\gamma\ j\ m')\ \mathfrak{D}^{(k)}_{q'q}(\omega)\ \mathfrak{D}^{(j)}_{m'm}(\omega)$$

Thus the vector $T(k\ q)u(\gamma\ j\ m)$ is transformed according to the product representation $\mathfrak{D}^{(k)} \otimes \mathfrak{D}^{(j)}$ of the rotation group, and hence may be expressed by use of vector-coupling coefficients as a linear combination of quantities each of which is transformed according to an irreducible representation.

Thus we get

$$T(k\ q)u(\gamma\ j\ m) = \sum_{j'm'} (k\ q\ j\ m|k\ j\ j'\ m')\Phi(j'\ m')$$

That is,

$$\Phi(j'\ m') = \sum_{q} T(k\ q)u(\gamma\ j\ m)(k\ q\ j\ m|k\ j\ j'\ m')$$

by the orthogonality of the V-C coefficients. The $\Phi(j'\ m')$ are simultaneously eigenvectors of $\mathbf{J}^2$ and J_z with eigenvalues j' and m'.

The matrix element of $T(kq)$ in the scheme $u(\gamma jm)$

$$(\gamma'\ j'\ m'|T(k\ q)|\gamma\ j\ m) \equiv (u(\gamma'\ j'\ m'),\ T(k\ q)u(\gamma\ j\ m))$$

is, due to the assumed orthonormality of the $u(\gamma jm)$, equal to

(5.3.1) $\qquad (u(\gamma'\ j'\ m'),\ \Phi(j'\ m'))(k\ q\ j\ m|k\ j\ j'\ m')$

We now prove a theorem which is used in the interpretation of the result (5.3.1).

THEOREM. Consider the transformation

$$v(\alpha\ j\ m) = \sum_{\beta} w(\beta\ j\ m)(\beta\ j\ m|\alpha\ j\ m)$$

Then (supposing $m < j$),

$$v(\alpha\ j\ m{+}1) = \sum_{\beta} w(\beta\ j\ m{+}1)(\beta\ j\ m{+}1|\alpha\ j\ m{+}1)$$

But

$$v(\alpha\ j\ m{+}1) = \frac{J_+}{\hbar[(j-m)(j+m+1)]^{\frac{1}{2}}} \cdot v(\alpha\ j\ m)$$
$$= \sum_{\beta} w(\beta\ j\ m{+}1)(\beta\ j\ m|\alpha\ j\ m)$$

from the original expression (cf. (2.3.16)). Hence

$$(\beta\ j\ m{+}1|\alpha\ j\ m{+}1) = (\beta\ j\ m|\alpha\ j\ m)$$

and

(5.3.2) *the transformation coefficients $(\beta j\ m|\alpha j\ m)$ are independent of m.*

Thus we see that the left-hand factor in (5.3.1) is independent of m; i.e. does not depend on the choice of orientation of the frame of reference. It is in fact determined solely by the physical properties of the operator and system. The geometrical or rotational dependence of the matrix element is concentrated in the right-hand factor, the vector-coupling coefficient.

This factorization is fundamental in the calculus of tensor operators, and is the basis of the great simplification of formulas which results from its use. The above theorem may be proved in a somewhat different way by starting with the definition (5.2.3) of the tensor operator in terms of the commutators with the components of angular momentum.

5.4. The Reduced Matrix Elements of a Tensor Operator

DEFINITION. It is convenient to define scalar quantities which differ slightly from the left-hand factor in (5.3.1). The $V\text{-}C$ coefficient is replaced by one more symmetrical in the quantum numbers of initial and final states, by use of the symmetry relation (3.7.4). We have then the definition of the *reduced* or *double-bar* matrix elements

(5.4.1)
$$\begin{aligned}
(\gamma'\ j'\ m'|T(k\ q)|\gamma\ j\ m) \\
= (-1)^{i-m}\frac{(j'\ m'\ j\ -m|j'\ j\ k\ q)}{(2k+1)^{\frac{1}{2}}}\ (\gamma'\ j'||T(k)||\gamma\ j) \\
= (-1)^{j'-m'}\begin{pmatrix} j' & k & j \\ -m' & q & m \end{pmatrix}(\gamma'\ j'||T(k)||\gamma\ j) \\
= (-1)^{k-j+j'}\frac{(k\ q\ j\ m|k\ j\ j'\ m')}{(2j'+1)^{\frac{1}{2}}}\ (\gamma'\ j'||T(k)||\gamma\ j).
\end{aligned}$$

The convention we have adopted is that of Racah (1942); it is compared with the conventions and notations used by some other workers in Table 5.1.

The orthogonality of the $V\text{-}C$ coefficients (3.5.4) enables us to write the alternative expression

(5.4.2)
$$\begin{aligned}
\delta_{kk'}\delta_{qq'}(\gamma'\ j'||T(k)||\gamma\ j) \\
= (2k+1)^{\frac{1}{2}} \sum_{mm'} (-1)^{j-m}(j'\ j\ k'\ q'|j'\ m'\ j\ -m) \\
\times (\gamma'\ j'\ m'|T(k\ q)|\gamma\ j\ m) \\
= (2k+1) \sum_{mm'} (-1)^{j'-m'}\begin{pmatrix} j' & k' & j \\ -m' & q' & m \end{pmatrix} \\
\times (\gamma'\ j'\ m'|T(k\ q)|\gamma\ j\ m) \\
= \sum_{mm'q} (-1)^{j'-m'}\begin{pmatrix} j' & k' & j \\ -m' & q' & m \end{pmatrix}(\gamma'\ j'\ m'|T(k\ q)|\gamma\ j\ m)
\end{aligned}$$

COMPUTATION OF REDUCED MATRIX ELEMENTS. The double-bar matrix elements are computed in practice in the obvious way; we choose the easiest to compute of the components $(\gamma' \, j' \, m'|T(k \, q)|\gamma \, j \, m)$ and divide it by

$$(-1)^{j'-m'}\begin{pmatrix} j' & k & j \\ -m' & q & m \end{pmatrix}$$

It is usually best to take $m' = m = q = 0$ or $m' = m = \frac{1}{2}, q = 0$, so that the simpler formulas of (3.7.17) or Table 2 may be employed.

We have for example

(5.4.3) $\qquad (l'||\mathbf{L}||l) = \hbar \delta_{l'l}[(2l + 1)(l + 1)l]^{\frac{1}{2}}$

and

(5.4.4) $\qquad (\frac{1}{2}||\mathbf{S}||\frac{1}{2}) = \hbar\sqrt{\frac{3}{2}}$

We use (4.6.3) to obtain the double-bar matrix elements for the spherical harmonics $Y_{kq}(\theta\varphi)$ where r, θ, φ are the particle coordinates. We get (cf. Racah (1942))

(5.4.5) $\quad (l'||\mathbf{Y}(k)||l) = (-1)^{l'}\left[\dfrac{(2l'+1)(2k+1)(2l+1)}{4\pi}\right]^{\frac{1}{2}}\begin{pmatrix} l' & k & l \\ 0 & 0 & 0 \end{pmatrix}$

(5.4.6) $\quad (l'||\mathbf{C}(k)||l) = (-1)^{l'}[(2l'+1)(2l+1)]^{\frac{1}{2}}\begin{pmatrix} l' & k & l \\ 0 & 0 & 0 \end{pmatrix}$

(For definition of $C_q^{(k)}$ see (2.5.31).)

TRANSITION PROBABILITIES. In the notation of Condon and Shortley (1935), p. 98, the *total intensity* of a line summed over the intensities of its components, i.e. over magnetic quantum numbers and polarizations is

(5.4.7) $\qquad S(\alpha \, j, \alpha' \, j') = \sum_{mqm'} |(\alpha \, j \, m|T(k \, q)|\alpha' \, j' \, m')|^2$

$\qquad\qquad\qquad = |(\alpha \, j||\mathbf{T}(k)||\alpha' \, j')|^2$

by (5.4.1) and (3.7.8), where $\mathbf{T}(k)$ is the operator inducing transitions.

APPROXIMATE EXPRESSIONS FOR LARGE j. The result in Appendix 2 may be used to give an approximate expression for the matrix element of a tensor operator between states of large angular momentum. We get

(5.4.8) $\qquad (\gamma' \, j+\delta \, m+q|T(k \, q)|\gamma \, j \, m) \cong d_{q\delta}^{(k)}(\theta) \cdot \dfrac{(\gamma' \, j+\delta||\mathbf{T}(k)||\gamma \, j)}{(2j+1)^{\frac{1}{2}}}$

where

$$\cos\theta = \dfrac{m}{\sqrt{j(j+1)}}$$

In the case of diagonal matrix elements[11] we see that the value represents, roughly speaking, the projection of the particular tensor component (considered as a c-number expression) on the direction of the angular momentum vector corresponding to j, m (see (2.7)). When $q = \delta = 0$ the d becomes $P_k(\cos\theta)$, and in particular[12]

$$(5.4.9) \qquad (\gamma\ j\ j|T(k\ 0)|\gamma\ j\ j) \cong \frac{(\gamma\ j||T(k)||\gamma\ j)}{(2j+1)^{\frac{1}{2}}}$$

The matrix element on the left clearly has the greatest magnitude of all the diagonal matrix elements of $T(k)$ for a given state γj. If $T(k)$ represents a multipole moment, then $(\gamma\ j\ j|T(k\ 0)|\gamma\ j\ j)$ is conventionally taken as the value of the moment for the state γj of the system (e.g. the magnetic moment or electric quadrupole moment of a nucleus[13]).

5.5. Hermitian Adjoint of Tensor Operators[14]

Let us take the Hermitian adjoint of the defining equation (5.2.1) We have

$$D(\omega)T(k\ q)^{\dagger}D^{-1}(\omega) = \sum_{q'} T(k\ q')^{\dagger} \mathfrak{D}^{(k)}_{q'q}{}^*(\omega).$$

Use of the symmetry property (4.2.7) of the $\mathfrak{D}$'s gives

$$D(\omega)T(k\ q)^{\dagger}D^{-1}(\omega) = \sum_{q'} T(k\ q')^{\dagger}(-1)^{q'-q}\mathfrak{D}^{(k)}_{-q'-q}(\omega)$$

I.e. the quantity $(-1)^q T(k\ -q)^{\dagger}$ transforms under rotations in the same way as $T(kq)$.

The concept of Hermitian adjoint of an operator may thus be generalized, and the Hermitian adjoint $T^{\dagger}$ of a tensor operator T may be defined by

$$(5.5.1) \qquad T^{\dagger}(k\ q) = (-1)^q (T(k\ -q))^{\dagger};$$

$T^{\dagger}$ clearly transforms under rotations like T; self-adjoint tensors $T^{\dagger} = T$ can only exist for integer k. Their components satisfy

$$(5.5.2) \qquad T(k\ q)^{\dagger} = (-1)^q T(k\ -q)$$

---

[11]See (4.1.24).
[12]Since

$$\begin{pmatrix} j & k & j \\ -j & 0 & j \end{pmatrix} = (2j+k+1)^{-\frac{1}{2}} \left[\frac{(2j)!(2j)!}{(2j+k)!(2j-k)!}\right]^{\frac{1}{2}}$$

from (3.7.11).
[13]Cf. Blatt and Weisskopf (1952) pp. 23–39.
[14]Cf. Racah (1942), Wigner (1951), Schwinger (1952).

This property is not conserved for tensor products; i.e. the tensor product of two self-adjoint tensor operators is not itself self-adjoint. However if we define a self-adjoint tensor operator by

(5.5.3) $$T(k\ q)^\dagger = (-1)^{k+q} T(k\ -q)$$

the property is conserved for tensor products (cf. (3.9)) provided the two tensors commute. Rose and Osborn (1954) discuss the problem of finding the Hermitian adjoint of tensor operators which do not commute. The property (5.5.2) is shared by tensor operators built up from Hermitian vector components V_x, V_y, V_z according to the formulas of Table 1 (see p. 124). These correspond to the phase convention with symmetry (2.5.6) for the Y_{lm} (example (5.1.3)). If on the other hand we multiply these quantities by i^k we obtain tensor operators with the symmetry (2.5.8) of the $\mathfrak{Y}_{lm}$, which have the property (5.5.3).

The reduced matrix elements must, as a result of the symmetry property of the V-C coefficients (see (5.4.1), (3.7.4), (3.7.5)) satisfy the condition

(5.5.4) $$(\gamma'\ j'||T(k)||\gamma\ j) = (-1)^{j'-j}(\gamma\ j||T^\dagger(k)||\gamma'\ j')^*$$

if we choose the definition (5.5.2). If we choose (5.5.3) we have

(5.5.5) $$(\gamma'\ j'||T(k)||\gamma\ j) = (-1)^{k+j'-j}(\gamma\ j||T^\dagger(k)||\gamma'\ j')^*$$

5.6. Electric Quadrupole Moment of Proton or Electron

We consider as an example of the foregoing the evaluation of the electric quadrupole moment of a proton or electron in a quantum state of definite angular momentum.

The classical quadrupole moment is

$$Q(p) = e \int (3z^2 - r^2)\rho(\mathbf{r})d^{(3)}\mathbf{r},$$

where $\rho(\mathbf{r})$ is the density of charge. In quantum mechanics the result is

$$Q(\rho) = e \int \psi^*(\mathbf{r})(3z^2 - r^2)\psi(\mathbf{r})d^{(3)}\mathbf{r}.$$

We define the tensor operator $\mathbf{Q}^{(2)}$ by

$$Q_0^{(2)} = +(3z^2 - r^2)$$

and take Q conventionally as[15]

$$Q = (j\ j|Q_0^{(2)}|j\ j) = (j||\mathbf{Q}^{(2)}||j)\begin{pmatrix} j & j & 2 \\ -j & j & 0 \end{pmatrix}$$

---
[15] Cf. Ramsay (1953) p. 361.

This is the quantity normally referred to as the quadrupole moment of the system. Now we compute the expectation value of the quadrupole moment in a state $j\,m$.

$$(j\,m|Q_0^{(2)}|j\,m) = (j||\mathbf{Q}^{(2)}||j)\begin{pmatrix} j & j & 2 \\ -m & m & 0 \end{pmatrix}$$

$$= (-1)^{j-m} Q \begin{pmatrix} j & j & 2 \\ -m & m & 0 \end{pmatrix} \Big/ \begin{pmatrix} j & j & 2 \\ -j & j & 0 \end{pmatrix}$$

$$= Q \cdot \frac{3m^2 - j(j+1)}{j(2j-1)}$$

by use of Table 2 (see p. 125), agreeing with the expression of Blatt and Weisskopf (1952) p. 28.

5.7. The Gradient Formula

The Wigner-Eckart theorem finds an important application in the derivation of the gradient formula,[16] which gives the gradient of a function F of the space coordinates when expressed in the form $F = \Phi(r) Y_{lm}(\theta, \varphi)$.

We evaluate first the matrix elements $(l'\,0|\nabla_0|l\,0)$ of the gradient operator, which is an example of a vector operator. We have

$$\nabla_0 = \frac{\partial}{\partial z} = \cos\theta \frac{\partial}{\partial r} - \frac{\sin\theta}{r}\frac{\partial}{\partial \theta}.$$

It is therefore necessary to evaluate the quantities $\cos\theta Y_{lm}(\theta, \varphi)$ and $\sin\theta \partial Y_{lm}(\theta, \varphi)/\partial\theta$. The properties of the Legendre functions (see (2.5.20), (2.5.25)) give us

$$\cos\theta Y_{lm}(\theta, \varphi) = \frac{l+1}{[(2l+1)(2l+3)]^{\frac{1}{2}}} Y_{l+1\,m}$$
$$+ \frac{l}{[(2l-1)(2l+1)]^{\frac{1}{2}}} Y_{l-1\,m}$$

$$\sin\theta \frac{\partial}{\partial \theta} Y_{lm}(\theta, \varphi) = \frac{l(l+1)}{[(2l+1)(2l+3)]^{\frac{1}{2}}} Y_{l+1\,m}$$
$$- \frac{l(l-1)}{[(2l-1)(2l+1)]^{\frac{1}{2}}} Y_{l-1\,m}$$

The only nonzero matrix elements of type $(l'\,0|\nabla_0|l\,0)$ are

$$(l+1\,0|\nabla_0|l\,0) = \frac{l+1}{[(2l+1)(2l+3)]^{\frac{1}{2}}} \left(\frac{\partial}{\partial r} - \frac{l}{r}\right)\Phi(r)$$

$$(l-1\,0|\nabla_0|l\,0) = \frac{l}{[(2l-1)(2l+1)]^{\frac{1}{2}}} \left(\frac{\partial}{\partial r} + \frac{l+1}{r}\right)\Phi(r)$$

---

[16] Cf. Bethe (1933) p. 558, Rose (1955) p. 24.

The general matrix elements are given by

$$(l'\ m'|\nabla_\mu|l\ m) = (-1)^{m'} \frac{\begin{pmatrix} l' & 1 & l \\ -m' & \mu & m \end{pmatrix}}{\begin{pmatrix} l' & 1 & l \\ 0 & 0 & 0 \end{pmatrix}} \cdot (l'\ 0|\nabla_0|l\ 0).$$

Evaluation of the 3-j symbols by use of Table 2 yields finally

$(l+1\ m+\mu|\nabla_\mu|l\ m)$

(5.7.1)
$$= \frac{(-1)^{l+m} A_\mu^+}{[2(2l+3)(2l+1)]^{\frac{1}{2}}} \left(\frac{\partial}{\partial r} - \frac{l}{r}\right)\Phi(r)$$

$$= \left(\frac{l+1}{2l+3}\right)^{\frac{1}{2}} (l\ m\ 1\ \mu|l\ 1\ l+1\ m+\mu) \left(\frac{\partial}{\partial r} - \frac{l}{r}\right)\Phi(r)$$

where

$$A_1^+ = [(l+m+1)(l+m+2)]^{\frac{1}{2}}$$
$$A_0^+ = -[2(l+m+1)(l-m+1)]^{\frac{1}{2}}$$
$$A_{-1}^+ = [(l-m+1)(l-m+2)]^{\frac{1}{2}}$$

$(l-1\ m+\mu|\nabla_\mu|l\ m)$

(5.7.2)
$$= \frac{(-1)^{l+m} A_\mu^-}{[2(2l+1)(2l-1)]^{\frac{1}{2}}} \left(\frac{\partial}{\partial r} + \frac{l+1}{r}\right)\Phi(r)$$

$$= -\left(\frac{l}{2l-1}\right)^{\frac{1}{2}} (l\ m\ 1\ \mu|l\ 1\ l-1\ m+\mu) \left(\frac{\partial}{\partial r} + \frac{l+1}{r}\right)\Phi(r)$$

where

$$A_1^- = [(l-m-1)(l-m)]^{\frac{1}{2}}$$
$$A_0^- = [2(l+m)(l-m)]^{\frac{1}{2}}$$
$$A_{-1}^- = [(l+m-1)(l+m)]^{\frac{1}{2}}$$

5.8. Expansion of a Plane Wave in Spherical Waves

We consider first a plane wave of wave number k moving in the direction of the positive z axis; it is given by

$$\exp(ikz) \equiv \exp(ikr\cos\theta)$$

Since the wave is symmetric about the z axis, its expansion in spherical waves can only contain $Y_{lm}(\theta, \varphi)$ with $m = 0$. The coefficients in the expansion are, as a result of the orthogonality of the spherical harmonics, (2.5.4)

$$c_{l0}(r) = 2\pi \int_0^\pi Y_{l0}(\theta) \exp(ikr\cos\theta) \sin\theta\ d\theta$$

which are expressed in terms of the Bessel functions of the first kind and half-odd-integer order (cf. Whittaker and Watson (1946) p. 398, Morse and Feshbach (1953) p. 1574). Thus we get

$$\exp(ikr \cos\theta) = \sum_{l=0}^{\infty} i^l \pi \left[\frac{2(2l+1)}{kr}\right]^{\frac{1}{2}} J_{l+\frac{1}{2}}(kr) Y_{l0}(\theta)$$
(5.8.1)
$$= \sum_{l=0}^{\infty} i^l [4\pi(2l+1)]^{\frac{1}{2}} j_l(kr) Y_{l0}(\theta)$$

where the spherical Bessel functions $j_l(z)$ are defined by

(5.8.2) $$j_l(z) = \left[\frac{\pi}{2z}\right]^{\frac{1}{2}} J_{l+\frac{1}{2}}(z)$$

The addition theorem (4.6.6) gives the expansion for a plane wave in an arbitrary direction Θ, Φ:

(5.8.3) $$\exp(i\mathbf{k}\cdot\mathbf{r}) = 4\pi \sum_{l=0}^{\infty} \sum_{m=-l}^{l} i^l j_l(kr) Y_{lm}(\theta,\varphi) Y_{lm}^*(\Theta,\Phi).$$

5.9. Vector Spherical Harmonics

The transformation under a rotation of the frame of reference of the functions representing a vector field is more complicated than for the case of a scalar field discussed in (4.1). We have to take into account the fact that the components of the vector field are defined with respect to the axes of the appropriate frame. A simple example will make the situation clear; we consider a rotation α about the z axis (represented by $D(\alpha\,0\,0)$). For convenience the Cartesian components of the vector field and the spherical polar coordinates of the field points are employed. The field is described in the original frame S by $V_x(r,\theta,\varphi)$, $V_y(r,\theta,\varphi)$ and $V_z(r,\theta,\varphi)$. In the new frame S''' obtained from S by the rotation $(\alpha\,0\,0)$ the components are

$$V_x'(r,\theta,\varphi) = \cos\alpha\, V_x(r,\theta,\varphi+\alpha) + \sin\alpha\, V_y(r,\theta,\varphi+\alpha)$$
$$V_y'(r,\theta,\varphi) = -\sin\alpha\, V_x(r,\theta,\varphi+\alpha) + \cos\alpha\, V_y(r,\theta,\varphi+\alpha)$$
$$V_z'(r,\theta,\varphi) = V_z(r,\theta,\varphi+\alpha)$$

The operator J_z of infinitesimal rotation about the z axis may be found by allowing α to become small. Thus we have

$$\mathbf{V}' = (1 + i\alpha J_z)\mathbf{V} + O(\alpha^2)$$

where

$$J_z = -i\left(x\frac{\partial}{\partial y} - y\frac{\partial}{\partial x}\right) + i\mathbf{e}_z \times$$

$\mathbf{e}_x$, $\mathbf{e}_y$ and $\mathbf{e}_z$ are the unit vectors along the x, y and z axes respectively and $\times$ indicates the vector product. Analogous results for the components J_x and J_y are easily obtained. The differential operators appearing on the right in the expressions for the components of $\mathbf{J}$ are recognized as (apart from a factor $\hbar$) the components of orbital angular momentum $\mathbf{L}$. (Cf. (2.1.3)). We omit this factor $\hbar$ since the present discussion is of a purely classical nature.

The components S_x, S_y and S_z of $\mathbf{S}$ are defined by

(5.9.1) $\qquad S_x = i\mathbf{e}_x \times; \qquad S_y = i\mathbf{e}_y \times; \qquad S_z = i\mathbf{e}_z \times$

and satisfy commutation relations similar to those of angular momentum operators. (Cf. Franz (1950))

Thus we have

(5.9.2) $\qquad\qquad\qquad \mathbf{J} = \mathbf{L} + \mathbf{S}$

The components of $\mathbf{L}$ commute with those of $\mathbf{S}$. The components of $\mathbf{J}$ have the important property that they commute with the *curl* operator:

(5.9.3) $\qquad J_\xi \nabla \times = \nabla \times J_\xi \qquad (\xi = x, y, \text{ or } z)$

Eigenvectors of $\mathbf{S}^2$ and S_z may now be found by taking suitable linear combinations of the unit vectors $\mathbf{e}_x$, $\mathbf{e}_y$ and $\mathbf{e}_z$. We define

(5.9.4)
$$\mathbf{e}_{+1} = -\frac{1}{\sqrt{2}}(\mathbf{e}_x + i\mathbf{e}_y)$$
$$\mathbf{e}_0 = \mathbf{e}_z$$
$$\mathbf{e}_{-1} = \frac{1}{\sqrt{2}}(\mathbf{e}_x - i\mathbf{e}_y)$$

and obtain

(5.9.5) $\qquad \mathbf{S}^2 \mathbf{e}_q = 2\mathbf{e}_q \qquad$ where $q = \pm 1, 0$
$\qquad\qquad\quad S_z \mathbf{e}_q = q\mathbf{e}_q$

Thus we have an angular momentum system with "spin" 1 (i.e. belonging to the representation $\mathfrak{D}^{(1)}$). The *spherical unit vectors* $\mathbf{e}_q$ have in addition the following properties:

The complex conjugate is given by

(5.9.6) $\qquad \mathbf{e}_q^* = (-1)^q \mathbf{e}_{-q} \qquad$ where $q = \pm 1, 0$

and the scalar product by

(5.9.7) $\qquad \mathbf{e}_q^* \cdot \mathbf{e}_{q'} = (-1)^q \mathbf{e}_q \cdot \mathbf{e}_{-q'} = \delta_{qq'}$

A vector quantity whose components are given in the spherical notation (5.1.3) appears as

(5.9.8) $\qquad\qquad\qquad \mathbf{V} = \sum_q (-1)^q V_q \mathbf{e}_{-q}$

and the components themselves may be expressed as scalar products

(5.9.9) $$V_q = \mathbf{e}_q \cdot \mathbf{V}$$

We now make use of the fact that (5.9.2) is an example of angular momentum addition (cf. Chap. 3) to evaluate the eigenvectors of $\mathbf{J}^2$ and J_z, that is, the *vector spherical harmonics*. The application of (3.5.1) gives

(5.9.10) $$\mathbf{Y}_{JlM}(\theta, \varphi) = \sum_{mq} Y_{lm}(\theta, \varphi) \mathbf{e}_q (l\ m\ 1\ q | l\ 1\ J\ M)$$

They have the properties

(5.9.11) $$\mathbf{J}^2 \mathbf{Y}_{JlM} = J(J+1) \mathbf{Y}_{JlM}$$

(5.9.12) $$J_z \mathbf{Y}_{JlM} = M \mathbf{Y}_{JlM}$$

(5.9.13) $$\int_0^{2\pi} \int_0^{\pi} \mathbf{Y}^*_{JlM}(\theta, \varphi) \cdot \mathbf{Y}_{J'l'M'}(\theta, \varphi) \sin\theta\, d\theta\, d\varphi = \delta_{JJ'} \delta_{ll'} \delta_{MM'}$$

As a result of the angular momentum addition rules, there are only three different types of vector spherical harmonics with given J, M. These three types divide into two classes from the point of view of parity; thus we have

$\mathbf{Y}_{JJM}$ with parity $(-1)^J$ corresponding to the magnetic field of *electric* multipole radiation and the electric field of *magnetic* multipole radiation.[17]

$\mathbf{Y}_{JJ\pm 1\ M}$ with parity $(-1)^{J+1}$ corresponding to the electric field of *electric* multipole radiation[17] and the magnetic field of *magnetic* multipole radiation.

The eigenvalues of $\mathbf{J}^2$ and J_z have an additional significance; it may be shown that a quantum with energy $\hbar\omega$ associated with a field represented by a vector spherical harmonic $\mathbf{Y}_{JlM}(\theta\varphi)$ has total angular momentum $\hbar\sqrt{J(J+1)}$ and the component of its angular momentum along the z axis is $\hbar M$. (cf. Franz (1950), Blatt and Weisskopf (1952).)

These vector spherical harmonics may be generated from scalar spherical harmonics by the use of certain operators; for example

(5.9.14) $$\mathbf{L} Y_{lm} = \sqrt{l(l+1)}\ \mathbf{Y}_{llm}$$

This result is proved by first writing the components of $\mathbf{L}$ in the spherical tensor notation; i.e. we have (cf. (2.3.1))

$$L_{+1} = -\frac{1}{\sqrt{2}} L_+, \qquad L_{-1} = \frac{1}{\sqrt{2}} L_-$$

---

[17] Cf. Blatt and Weisskopf (1952) p. 799, Franz (1950).

84   5 · SPHERICAL TENSORS AND TENSOR OPERATORS

It follows from (2.3.16), (2.3.17) and Table 5.2 that we may write

(5.9.15) $\quad L_q Y_{lm} = (-1)^q \sqrt{l(l+1)}\, Y_{lm+q}(l\,m+q\,1,\,-q|l\,1\,l\,m)$

Reference to (5.9.4) and (5.9.10) gives immediately the required expression (5.9.14).

We may also obtain vector spherical harmonics by application of the unit vector $\mathbf{r}/r$:

(5.9.16) $\quad \dfrac{\mathbf{r}}{r} Y_{lm} = -\left[\dfrac{l+1}{2l+1}\right]^{\frac{1}{2}} \mathbf{Y}_{l\,l+1\,m} + \left[\dfrac{l}{2l+1}\right]^{\frac{1}{2}} \mathbf{Y}_{l\,l-1\,m}$

The gradient formula ((5.7.1), (5.7.2)) may be expressed in vector form:

(5.9.17) $\quad \nabla \Phi(r) Y_{lm} = -\left(\dfrac{l+1}{2l+1}\right)^{\frac{1}{2}} \left(\dfrac{d}{dr} - \dfrac{l}{r}\right) \Phi \cdot \mathbf{Y}_{l\,l+1\,m}$
$\qquad\qquad\qquad + \left(\dfrac{l}{2l+1}\right)^{\frac{1}{2}} \left(\dfrac{d}{dr} + \dfrac{l+1}{r}\right) \Phi \cdot \mathbf{Y}_{l\,l-1\,m}$

Since the operator $\nabla \times$ commutes with the components of $\mathbf{J}$, the curl of a vector spherical harmonic is a linear combination of vector spherical harmonics with the same J and M. An arbitrary (sufficiently well-behaved) vector field may be expressed as a series of multipole fields each consisting of the product of a function of r and a vector spherical harmonic; the result of applying the curl operator to the three typical products of this kind is now given.[18]

(5.9.18) $\quad \nabla \times (\Phi(r) \mathbf{Y}_{l\,l+1\,M}(\theta,\varphi)) = i\left(\dfrac{d}{dr} + \dfrac{l+2}{r}\right) \Phi \cdot \left(\dfrac{l}{2l+1}\right)^{\frac{1}{2}} \mathbf{Y}_{l\,l\,M}$

(5.9.19) $\quad \nabla \times (\Phi(r) \mathbf{Y}_{l\,l\,M}(\theta,\varphi)) = i\left(\dfrac{d}{dr} - \dfrac{l}{r}\right) \Phi \cdot \left(\dfrac{l}{2l+1}\right)^{\frac{1}{2}} \mathbf{Y}_{l\,l+1\,M}$
$\qquad\qquad\qquad + i\left(\dfrac{d}{dr} + \dfrac{l+1}{r}\right) \Phi \cdot \left(\dfrac{l+1}{2l+1}\right)^{\frac{1}{2}} \mathbf{Y}_{l\,l-1\,M}$

(5.9.20) $\quad \nabla \times (\Phi(r) \mathbf{Y}_{l\,l-1\,M}(\theta,\varphi)) = i\left(\dfrac{d}{dr} - \dfrac{l-1}{r}\right) \Phi \cdot \left(\dfrac{l+1}{2l+1}\right)^{\frac{1}{2}} \mathbf{Y}_{l\,l\,M}$

Note that the curl operator changes the parity of the function.

The divergence operator gives the following scalar quantities:

(5.9.21) $\quad \nabla \cdot (\Phi(r) \mathbf{Y}_{l\,l+1\,M}(\theta,\varphi)) = -\left(\dfrac{l+1}{2l+1}\right)^{\frac{1}{2}} \left(\dfrac{d}{dr} + \dfrac{l+2}{r}\right) \Phi \cdot Y_{lM}$

(5.9.22) $\quad \nabla \cdot (\Phi(r) \mathbf{Y}_{l\,l\,M}(\theta,\varphi)) = 0 \quad \text{for any } \Phi(r)$

---

[18] The six expressions (5.9.18)–(5.9.23) are due to fil. lic. P. O. Olsson.

$$(5.9.23) \quad \nabla \cdot (\Phi(r) \mathbf{Y}_{l\,l-1\,M}(\theta, \varphi)) = \left(\frac{l}{2l+1}\right)^{\frac{1}{2}} \left(\frac{d}{dr} - \frac{l-1}{r}\right) \Phi \cdot Y_{lM}$$

The application of vector spherical harmonics to problems of the electromagnetic field is discussed by Blatt and Weisskopf (1952) and by Franz (1950). A different and more fundamental derivation of some of the above results is given by Corben and Schwinger (1940).

5.10. Spin Spherical Harmonics

These are defined in a similar way to the vector spherical harmonics; in place of a vector field we have a spinor field. The operator of total angular momentum is given by

$$\mathbf{J} = \mathbf{L} + \mathbf{s}$$

The operator $\mathbf{s}$ is associated with the representation $\mathfrak{D}^{(\frac{1}{2})}$ of the group of rotations, i.e. with transformations in spinor space.

The eigenvectors of $\mathbf{J}^2$ and J_z are thus given by

$$(5.10.1) \quad \Phi_{JlM} = \sum_{m\mu} \mathfrak{Y}_{lm} \chi_\mu (l\,m\,\tfrac{1}{2}\,\mu | l\,\tfrac{1}{2}\,J\,M)$$

The choice of phase for the spherical harmonics (see (2.5.8)) gives convenient properties for Φ with respect to time reversal (cf. (3.9) and (5.11) and Biedenharn and Rose (1953) p. 736).

5.11. Emission and Absorption of Particles

We consider the transition probability for the absorption or emission of a particle by a system (say, a nucleus). The incident (or emitted) particle is represented by the state vector $u(\gamma_1 j_1 m_1)$, the target (or remaining) particle by $u(\gamma_b j_b m_b)$, and the product (or initial) system by $u(\gamma_a j_a m_a)$. The transition probability is given by the square modulus of such a matrix element as

$$(5.11.1) \quad (\gamma_1 j_1 m_1 \gamma_b j_b m_b | R | \gamma_a j_a m_a)$$
$$= \left(u(\gamma_1 j_1 m_1) u(\gamma_b j_b m_b), R u(\gamma_a j_a m_a)\right)$$

where the operator R is defined by $1 + R = S$, where S is the scattering matrix of Heisenberg. In a first order perturbation treatment R is proportional to the part H' of the Hamiltonian inducing the perturbation.

Now we may express the product $u(\gamma_1 j_1 m_1) u(\gamma_b j_b m_b)$ representing the separated parts of the system as a sum of terms each corresponding to definite eigenvalues of the operators $\mathbf{J}_a^2$, J_{az} of the total angular momentum.

(5.11.2)
$$u(\gamma_1\ j_1\ m_1)u(\gamma_b\ j_b\ m_b) = \sum_{j_a m_a} v(\gamma_1\ \gamma_b\ j_1\ j_b\ j_a\ m_a)(j_1\ j_b\ j_a\ m_a|j_1\ m_1\ j_b\ m_b)$$

where

(5.11.3)
$$v(\gamma_1\ \gamma_b\ j_1\ j_b\ j_a\ m_a) = \sum_{m_1 m_b} u(\gamma_1\ j_1\ m_1)u(\gamma_b\ j_b\ m_b)(j_1\ m_1\ j_b\ m_b|j_1\ j_b\ j_a\ m_a)$$

Now since R is a scalar operator, it is diagonal in $j_a\ m_a$; hence we have

$$(\gamma_1\ j_1\ m_1\ \gamma_b\ j_b\ m_b|R|\gamma_a\ j_a\ m_a)$$
$$= \Big(v(\gamma_1\ \gamma_b\ j_1\ j_b\ j_a\ m_a),\ Ru(\gamma_a\ j_a\ m_a)\Big) \cdot (j_1\ j_b\ j_a\ m_a|j_1\ m_1\ j_b\ m_b).$$

It follows from (5.3.2) that the first factor on the right is independent of m_a, i.e. is a scalar quantity; we may therefore write

(5.11.4)
$$(\gamma_1\ j_1\ m_1\ \gamma_b\ j_b\ m_b|R|\gamma_a\ j_a\ m_a)$$
$$= (\gamma_1\ \gamma_b\ j_1\ j_b||R||\gamma_a\ j_a)(j_1\ j_b\ j_a\ m_a|j_1\ m_1\ j_b\ m_b)$$

where the scalar quantity $(\gamma_1\ \gamma_b\ j_1\ j_b||R||\gamma_a\ j_a)$ is called the *reduced matrix element* of R; the reader should be careful to distinguish the reduced matrix element of this scalar operator from that of the component of a tensor operator (5.4.1).

THE REALITY OF THE MATRIX ELEMENTS OF R^{19}. We now investigate the conditions which must be satisfied if the matrix elements of R are to be real. Use is made of the operation K of time reversal (cf. (2.8)) which commutes with R.

The operator K may be represented in general by

$$K = UK_0$$

where U is a unitary matrix and K_0 is the operation of taking the complex conjugate. If we consider states which are eigenvectors of orbital angular momentum (j = integer) then U is the unit matrix 1. In that case

$$(Kv, KRu) = (Kv, RKu) = (v, Ru)^*$$

We shall see that if the angular momentum eigenvectors $u(\gamma\ j\ m)$ have the property under time reversal

(5.11.5)
$$Ku(\gamma\ j\ m) = (-1)^{j+m}u(\gamma\ j\ -m)$$

---

[19]Cf. Biedenharn and Rose (1953).

then the matrix elements of R are real. For we have

$$\left(Ku(\gamma_1 \ j_1 \ m_1)Ku(\gamma_b \ j_b \ m_b), RKu(\gamma_a \ j_a \ m_a)\right)$$

$$= (\gamma_1 \ \gamma_b \ j_1 \ j_b||R||\gamma_a \ j_a)^*(j_1 \ j_b \ j_a \ m_a|j_1 \ m_1 \ j_b \ m_b)$$

which is, if the condition (5.11.5) is satisfied by the angular momentum eigenvectors, equal to

$$\left(u(\gamma_1 \ j_1 \ -m_1)u(\gamma_b \ j_b \ -m_b), Ru(\gamma_a \ j_a \ -m_a)\right) \cdot (-1)^{j_1+m_1+j_b+m_b-j_a-m_a}$$

$$= (\gamma_1 \ \gamma_b \ j_1 \ j_b||R||\gamma_a \ j_a)(j_1 \ j_b \ j_a \ m_a|j_1 \ m_1 \ j_b \ m_b)$$

by use of the symmetry property of the V-C coefficient (3.5.17). It follows that the matrix elements of R and also the reduced matrix elements $(\gamma_1 \ \gamma_b \ j_1 \ j_b||R||\gamma_a \ j_a)$ are real.

Thus if we choose the eigenfunctions of orbital angular momentum with the Condon and Shortley phase, namely the Y_{lm}, their property under complex conjugation (2.5.6) makes it impossible to guarantee the reality of the matrix elements of R. On the other hand, if we take the eigenfunctions $\mathfrak{Y}_{lm}$ with the property (2.5.8), and if necessary construct eigenvectors of half-odd-integer angular momentum by compounding them with the spin eigenvectors which transform according to (2.8.2) under time reversal, then the condition (5.11.5) is satisfied and the matrix elements $(\gamma_1 \ j_1 \ m_1 \ \gamma_b \ j_b \ m_b|R|\gamma_a \ j_a \ m_a)$ are always real (see also (3.9)).

Table 5.1. Notations relating to tensor operators and reduced matrix elements. We assume throughout the usual convention, that

$$(a|Q|b) = \int \psi_a^* Q \psi_b \, d\tau$$

The q component of a tensor operator of rank k is written in these notes as $T(kq)$, and the matrix elements between the states $\alpha j m$ and $\alpha' j' m'$ as $(\alpha \ j \ m|T(kq)|\alpha' \ j' \ m')$. This is identical with the notation of Schwinger (1952). Other notations, which are equivalent to those already mentioned, are given below.

Racah (1942): q component of tensor operator of rank k : $T_q^{(k)}$.
  Matrix element of this operator: $(\alpha \ j \ m|T_q^{(k)}|\alpha' \ j' \ m')$

Wigner (1951): τ component of tensor operator of rank t : t_τ.
  Matrix element of this operator between states with angular momenta $l \ \kappa, \ l' \lambda : (\psi_\kappa^l, t_\tau \ \psi_\lambda^{l'})$.

Landau and Lifschitz (1948): m component of tensor operator of rank j : $f^{(jm)}$.
  Matrix element of this operator between states with angular momenta $j_1 m_1, \ j_2 m_2 : (f^{(jm)})_{j_2 m_2}^{j_1 m_1}$

Table 5.1 (Continued)

Biedenharn and Rose (1953): μ component of tensor operator of rank L, with parity $\pi(=\pm 1) : T(L\mu, \pi)$.
Matrix element of this operator between states with angular momenta $j_1 m_1, jm : \langle j_1 m_1 | T(L\mu, \pi) | jm \rangle$.

Fano (1951): q component of tensor operator of rank $k : T_{kq}$.
Matrix element of this operator between states with angular momenta $j'm'$ and $jm : \langle j' m' | T_{kq} | j\, m \rangle$.

The *double-bar* matrix element, or reduced matrix element, is defined in these notes as follows:

$$(\alpha\ j\ m | T(kq) | \alpha'\ j'\ m') = (-1)^{j-m} (\alpha\ j || T(k) || \alpha'\ j') \begin{pmatrix} j & k & j' \\ -m & q & m' \end{pmatrix}$$

This is identical with the definition of Racah (1942); however in his notation the relation is:

Racah (1942)

$$(\alpha\ j\ m | T_q^{(k)} | \alpha'\ j'\ m') = (-1)^{j+m} (\alpha\ j || T^{(k)} || \alpha'\ j') \times V(j\ j'k; -mm'\ q)$$

The definition of *Wigner* (1951) is also equivalent to ours:

$$(\psi_\kappa^l, t_\tau \psi_\lambda^{l'}) = (-1)^{l-\kappa} \begin{pmatrix} l & t & l' \\ -\kappa & \tau & \lambda \end{pmatrix} t_{ll'}$$

Other notations, which are *not* equivalent to ours, are given below; reference should be made to Table 3.1 for the notations for Clebsch-Gordan (V-C) coefficients.

Schwinger (1952):

$$(\gamma\ j\ m | T(kq) | \gamma'\ j'\ m') = (-1)^{k-j'+m} [\gamma\ j | T^{(k)} | \gamma'\ j'] X(j\ k\ j'; -m\ q\ m')$$

Biedenharn and Rose (1953):

$$\langle j_1\ m_1 | T(L\ \mu, \pi) | j\ m \rangle = C(j_1\ L\ j; m_1\ m - m_1)(j_1 || T(L\pi) || j) \delta(\pi, \pi_1\ \pi_a)$$

Landau and Lifschitz (1948):

$$(f^{(jm)})^{j_1 m_1}_{j_2 m_2} = (f^{(j)})^{j_1}_{j_2}(-1)^{m_2}\sqrt{2j_2+1}\ S_{j_1 m_1; jm; j_2, -m_2}$$

Fano (1951):

$$\sum_{m'm} \langle (j'\ j)\bar{k}\bar{q} | j'\ m', j\ -m\rangle (-1)^{j-m} \langle j'\ m' | T_{kq} | j\ m \rangle = \langle j'\ j || T_k \rangle \times \delta_{\bar{k}k}\delta_{\bar{q}q}$$

Condon and Shortley (1935): We relate the analogous notation of TAS for vector operators to our own by quoting equations (30) of Racah (1942):

$$(\alpha\ j || T^{(1)} || \alpha'\ j) = [j(j+1)(2j+1)]^{\frac{1}{2}} (\alpha\ j \vdots T \vdots \alpha'\ j)$$

$$(\alpha\ j || T^{(1)} || \alpha'\ j{-}1) = [j(2j-1)(2j+1)]^{\frac{1}{2}} (\alpha\ j \vdots T \vdots \alpha'\ j{-}1)$$

$$(\alpha\ j || T^{(1)} || \alpha'\ j{+}1) = -[(j+1)(2j+1)(2j+3)]^{\frac{1}{2}} (\alpha\ j \vdots T \vdots \alpha'\ j{+}1)$$

Table 5.2

$(j_1\, m_1\, \tfrac{1}{2}\, m_2 | j_1\, \tfrac{1}{2}\, j\, m)$

j \ m_2	$\tfrac{1}{2}$	$-\tfrac{1}{2}$
$j_1 + \tfrac{1}{2}$	$\left[\dfrac{j_1 + m + \tfrac{1}{2}}{2j_1 + 1}\right]^{\tfrac{1}{2}}$	$\left[\dfrac{j_1 - m + \tfrac{1}{2}}{2j_1 + 1}\right]^{\tfrac{1}{2}}$
$j_1 - \tfrac{1}{2}$	$-\left[\dfrac{j_1 - m + \tfrac{1}{2}}{2j_1 + 1}\right]^{\tfrac{1}{2}}$	$\left[\dfrac{j_1 + m + \tfrac{1}{2}}{2j_1 + 1}\right]^{\tfrac{1}{2}}$

$(j_1\, m_1\, 1\, m_2 | j_1\, 1\, j\, m)$

j \ m_2	1	0	-1
$j_1 + 1$	$\left[\dfrac{(j_1 + m)(j_1 + m + 1)}{(2j_1 + 1)(2j_1 + 2)}\right]^{\tfrac{1}{2}}$	$\left[\dfrac{(j_1 - m + 1)(j_1 + m + 1)}{(2j_1 + 1)(j_1 + 1)}\right]^{\tfrac{1}{2}}$	$\left[\dfrac{(j_1 - m)(j_1 - m + 1)}{(2j_1 + 1)(2j_1 + 2)}\right]^{\tfrac{1}{2}}$
j_1	$-\left[\dfrac{(j_1 + m)(j_1 - m + 1)}{2j_1(j_1 + 1)}\right]^{\tfrac{1}{2}}$	$\dfrac{m}{[j_1(j_1 + 1)]^{\tfrac{1}{2}}}$	$\left[\dfrac{(j_1 - m)(j_1 + m + 1)}{2j_1(j_1 + 1)}\right]^{\tfrac{1}{2}}$
$j_1 - 1$	$\left[\dfrac{(j_1 - m)(j_1 - m + 1)}{2j_1(2j_1 + 1)}\right]^{\tfrac{1}{2}}$	$-\left[\dfrac{(j_1 - m)(j_1 + m)}{j_1(2j_1 + 1)}\right]^{\tfrac{1}{2}}$	$\left[\dfrac{(j_1 + m + 1)(j_1 + m)}{2j_1(2j_1 + 1)}\right]^{\tfrac{1}{2}}$

CHAPTER 6

The Construction of Invariants from the Vector-Coupling Coefficients

6.1. The Recoupling of Three Angular Momenta

THE TWO COUPLING SCHEMES. Let us consider[1] the coupling of three angular momenta j_1, j_2, j_3 to give a resultant J. We note first that there is no unique way to carry out this coupling; we might (I) first couple j_1 and j_2 to give a resultant j_{12}, and couple this to j_3 to give J, or alternatively (II) couple j_1 to the resultant j_{23} of coupling j_2 and j_3. We remember also that the *order* of coupling determines the phase of the resulting state vector (see (3.5.14)).

If we choose some definite way to carry out the coupling, we find that in general there are several values of intermediate angular momentum (say j_{12}) which give a particular final J. Suppose for example we couple $j_1 = 1$ and $j_2 = 2$. The possible values of j_{12} are 1, 2, and 3. Then if $j_3 = 1$ and we require a final $J = 3$, either $j_{12} = 2$ or $j_{12} = 3$ can give this resultant when coupled to j_3. The states obtained with particular values of J and M but different intermediate values of j_{12} are independent, and must be distinguished by specification of the intermediate j values and the mode of coupling. Thus for a given JM we have in general a system of states, and they are represented in different ways by different modes of coupling; it follows that there exists a unitary transformation connecting two such representations.

Let us denote the state vector arising from a type I coupling by $v((j_1 j_2)j_{12}, j_3, JM)$ and that from a type II coupling by $v(j_1, (j_2 j_3)j_{23}, JM)$. These state vectors are given respectively by the equations

$$w((j_1\ j_2)j_{12},\ j_3,\ JM)$$

(6.1.1)
$$= \sum_{m_{12} m_3} v(j_1\ j_2\ j_{12}\ m_{12})u(j_3\ m_3)(j_{12}\ m_{12}\ j_3\ m_3|j_{12}\ j_3\ J\ M)$$

$$= \sum_{m_1 m_2 m_3 m_{12}} u(j_1\ m_1)u(j_2\ m_2)u(j_3\ m_3)$$

$$\times (j_1\ m_1\ j_2\ m_2|j_1\ j_2\ j_{12}\ m_{12})(j_{12}\ m_{12}\ j_3\ m_3|j_{12}\ j_3\ J\ M)$$

---

[1]Cf. for example Racah (1943).

6.1 · RECOUPLING OF THREE ANGULAR MOMENTA

and

(6.1.2)
$$w(j_1, (j_2\ j_3)j_{23}, J\ M)$$
$$= \sum_{m_1 m_{23}} u(j_1\ m_1)v(j_2\ j_3\ j_{23}\ m_{23})(j_1\ m_1\ j_{23}\ m_{23}|j_1\ j_{23}\ J\ M)$$
$$= \sum_{m_1 m_2 m_3 m_{23}} u(j_1\ m_1)u(j_2\ m_2)u(j_3\ m_3)$$
$$\times (j_2\ m_2\ j_3\ m_3|j_2\ j_3\ j_{23}\ m_{23})(j_1\ m_1\ j_{23}\ m_{23}|j_1\ j_{23}\ J\ M)$$

The unitary transformation connecting these two representations is given by

(6.1.3)
$$w(j_1, (j_2\ j_3)j_{23}, J\ M) = \sum_{j_{12}} w((j_1\ j_2)j_{12}, j_3, J\ M)$$
$$\times ((j_1\ j_2)j_{12}, j_3, J\ M|j_1, (j_2\ j_3)j_{23}, J\ M)$$

We see from (5.3.2) that the transformation coefficients are independent of M.

EVALUATION OF THE RECOUPLING COEFFICIENTS. We now make use of the orthogonality of the vector-coupling coefficients (3.5.3) and (3.5.4) to obtain

(6.1.4)
$$((j_1\ j_2)j_{12}, j_3, J|j_1, (j_2\ j_3)j_{23}, J)$$
$$= \sum_{\substack{m_1 m_2 m_3 \\ m_{12} m_{23}}} (j_{12}\ j_3\ J\ M|j_{12}\ m_{12}\ j_3\ m_3)(j_1\ j_2\ j_{12}\ m_{12}|j_1\ m_1\ j_2\ m_2)$$
$$\times (j_2\ m_2\ j_3\ m_3|j_2\ j_3\ j_{23}\ m_{23})(j_1\ m_1\ j_{23}\ m_{23}|j_1\ j_{23}\ J\ M)$$

where we have dropped the argument M in the transformation coefficient as a result of (5.3.2). The M appearing in the V-C coefficients must of course lie between $-J$ and J. Other transformation coefficients arising from the recoupling of three angular momenta will clearly differ only in a trivial way from the form just discussed; to evaluate them we need only make use of the rule for the changing of the order of coupling.

This transformation coefficient which we have just evaluated is of great importance in quantum mechanical problems; for we find that we often have to deal with the addition of a number of angular momenta, involving the summation of products of vector-coupling coefficients, the sum being over the magnetic quantum numbers m. Now the vector-coupling coefficients are not invariant under rotations of the frame of reference, while the quantities which we wish to compute —such as energies, cross-sections, transition probabilities, etc.—are usually scalars. Hence the V-C coefficients are associated in such a way that they form scalar quantities, which are functions only of the j values and not of the m's. We shall see how we may evaluate directly such scalar quantities as the transformation coefficient just derived, eliminating the tedious computation of masses of V-C coefficients.

6 · CONSTRUCTION OF INVARIANTS

DEFINITION OF THE 6-j SYMBOL. We shall now define a quantity associated with such transformations between coupling schemes of 3 angular momenta, namely the 6-j symbol.[2] The choice of normalization is such as to give the symbol the maximum symmetry with respect to permutations of its arguments. We define thus

(6.1.5)
$$\begin{Bmatrix} j_1 & j_2 & j_{12} \\ j_3 & J & j_{23} \end{Bmatrix} = [(2j_{12}+1)(2j_{23}+1)]^{-\frac{1}{2}} \cdot (-1)^{j_1+j_2+j_3+J}$$
$$\times ((j_1 \; j_2)j_{12}, \; j_3, \; J | j_1, \; (j_2 \; j_3)j_{23}, \; J)$$
$$= [(2j_{12}+1)(2j_{23}+1)]^{-\frac{1}{2}} \cdot (-1)^{j_1+j_2+j_3+J}$$
$$\times \sum_{m_1 m_2} (j_1 \; m_1 \; j_2 \; m_2 | j_1 \; j_2 \; j_{12} \; m_1+m_2)$$
$$\times (j_{12} \; m_1+m_2 \; j_3 \; M-m_1-m_2 | j_{12} \; j_3 \; J \; M)$$
$$\times (j_2 \; m_2 \; j_3 \; M-m_1-m_2 | j_2 \; j_3 \; j_{23} \; M-m_1)$$
$$\times (j_1 \; m_1 \; j_{23} \; M-m_1 | j_1 \; j_{23} \; J \; M)$$

where we have taken advantage of the rule $m_1 + m_2 = M$ to reduce the number of indices of summation.

The 6-j symbol is of importance in all situations where recoupling of angular momenta is involved; even when there are more than three angular momenta the invariant quantities (i.e. the recoupling coefficients) arising may be expressed in terms of 6-j symbols. A detailed investigation of the properties of this quantity is therefore justified.

6.2. The Properties of the 6-j Symbol

EVALUATION OF THE 6-j SYMBOL IN TERMS OF THE 3-j SYMBOLS. We shall now discuss the 6-j symbol from another point of view, with the purpose of clarifying its symmetry properties. To do this we make use of the 3-j symbol instead of the ordinary V-C coefficient. We remember that the 3-j symbol is associated with the coupling of three angular momenta to give zero resultant (a process which, contrary to those already discussed, can only be carried out in one way). Thus we may say that the expression

(6.2.1) $$\sum_{m_1 m_2 m_3} u(j_1 \; m_1) u(j_2 \; m_2) u(j_3 \; m_3) \begin{pmatrix} j_1 & j_2 & j_3 \\ m_1 & m_2 & m_3 \end{pmatrix}$$

is a scalar; it follows that the set of $(2j_1+1)(2j_2+1)(2j_3+1)$ 3-j symbols with given values of j_1, j_2, and j_3 and all possible corresponding values of m_1, m_2, and m_3 may be regarded as a *tensor* which transforms under rotations contragrediently to the set of products $u(j_1 \; m_1)u(j_2 \; m_2)u(j_3 \; m_3)$.

---

[2] Wigner (1951).

6.2 · PROPERTIES OF THE 6-j SYMBOL

Now we have already mentioned the possibility of defining a *metric tensor* $\begin{pmatrix} j \\ m\ m' \end{pmatrix}$ (cf. (3.7.1)); we may also define a corresponding *contraction process*. In principle we should make use of the contragredient metric tensor to $\begin{pmatrix} j \\ m\ m' \end{pmatrix}$; however, this is easily shown to be identical with $\begin{pmatrix} j \\ m\ m' \end{pmatrix}$. We shall carry out contractions of the indices (the magnetic quantum numbers) in products of 3-j symbols. We must remember that the contractions may only occur between 3-j symbols which contain the same j values. Thus the basic contraction process is exemplified by the expression

$$(6.2.2) \quad \sum_{m_3 m_3'} \begin{pmatrix} j_1 & j_2 & j_3 \\ m_1 & m_2 & m_3 \end{pmatrix} \begin{pmatrix} j_4 & j_3 & j_5 \\ m_4 & m_3' & m_5 \end{pmatrix} \begin{pmatrix} j_3 \\ m_3 & m_3' \end{pmatrix}$$

The question now is, what is the simplest nontrivial combination of products of 3-j symbols in which contractions may be carried out to give a resultant scalar? We represent a 3-j symbol by a point which is the vertex of three lines, each of which represents a j value. Each of the j values must be contracted with a similar j value from another 3-j symbol, i.e. each line must terminate at another vertex. It is clear that the simplest non-trivial diagram satisfying these conditions is a tetrahedron. That is, we may make a sum of products of four 3-j symbols, which contain all together six different j values, the six metric tensors being included so that a scalar quantity is the result. Let us then draw a tetrahedron (Fig. 6.1) and associate with each vertex a 3-j symbol

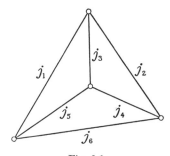

Fig. 6.1

and with each edge a j value. The three j values of each 3-j symbol are the j values of the edges meeting at the corresponding vertex.

We may construct an alternative diagram, in which the j values associated with each 3-j symbol occupy the edges of a face (Fig. 6.2). Since the 3-j symbols are only nonzero when the corresponding j values

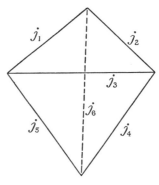

Fig. 6.2

form triangles, Fig. 6.2 has a metrical significance; the quantity we are constructing is only nonzero when the six j values chosen correspond to the lengths of the sides of a tetrahedron. This type of diagram is, however, of no use when we come to consider the 9-j symbol.

We choose a definite convention for carrying out the contraction process; we remember the symmetry property of the 3-j symbols (3.7.4), (3.7.5) and that the metric tensor is skew-symmetric for half-odd integer j.

$$
(6.2.3) \quad \begin{Bmatrix} j_1 & j_2 & j_3 \\ j_4 & j_5 & j_6 \end{Bmatrix}
= \sum_{\text{all } m} \begin{pmatrix} j_1 & j_2 & j_3 \\ m_1 & m_2 & m_3 \end{pmatrix} \begin{pmatrix} j_1 & j_5 & j_6 \\ m'_1 & m_5 & m'_6 \end{pmatrix} \begin{pmatrix} j_4 & j_2 & j_6 \\ m'_4 & m'_2 & m_6 \end{pmatrix} \begin{pmatrix} j_4 & j_5 & j_3 \\ m_4 & m'_5 & m'_3 \end{pmatrix}
$$
$$
\times \begin{pmatrix} j_1 \\ m_1 & m'_1 \end{pmatrix} \begin{pmatrix} j_2 \\ m_2 & m'_2 \end{pmatrix} \begin{pmatrix} j_3 \\ m_3 & m'_3 \end{pmatrix} \begin{pmatrix} j_4 \\ m_4 & m'_4 \end{pmatrix} \begin{pmatrix} j_5 \\ m_5 & m'_5 \end{pmatrix} \begin{pmatrix} j_6 \\ m_6 & m'_6 \end{pmatrix}
$$

If we rewrite this expression using V-C coefficients making use of (3.7.3) we may bring it into a form equivalent to (6.1.5). We note that one of the indices of summation is free; the summation over six indices is replaced by a summation over two, since we have the rule $m_1 + m_2 + m_3 = 0$.

SYMMETRIES OF THE 6-j SYMBOL.* The form into which we have cast the 6-j symbol makes it a simple matter to derive its symmetry properties. It is clearly left invariant by any permutation of the columns;

$$
(6.2.4) \quad \begin{Bmatrix} j_1 & j_2 & j_3 \\ j_4 & j_5 & j_6 \end{Bmatrix} = \begin{Bmatrix} j_2 & j_3 & j_1 \\ j_5 & j_6 & j_4 \end{Bmatrix} = \begin{Bmatrix} j_3 & j_1 & j_2 \\ j_6 & j_4 & j_5 \end{Bmatrix}
$$
$$
= \begin{Bmatrix} j_2 & j_1 & j_3 \\ j_5 & j_4 & j_6 \end{Bmatrix} = \begin{Bmatrix} j_1 & j_3 & j_2 \\ j_4 & j_6 & j_5 \end{Bmatrix} = \begin{Bmatrix} j_3 & j_2 & j_1 \\ j_6 & j_5 & j_4 \end{Bmatrix}
$$

---

*See also Jahn and Howell (1959).

The 6-j symbol is also invariant against interchange of the upper and lower arguments in each of any two columns. E.g.

(6.2.5)
$$\begin{Bmatrix} j_1 & j_2 & j_3 \\ j_4 & j_5 & j_6 \end{Bmatrix} = \begin{Bmatrix} j_1 & j_5 & j_6 \\ j_4 & j_2 & j_3 \end{Bmatrix}$$

In fact there are 24 operations generated by permutations of type (6.2.4) or (6.2.5) which leave a 6-j symbol invariant, and these form a group isomorphic with the symmetry group of a regular tetrahedron. Any of these operations corresponds to a rotation and/or reflection of the tetrahedron whose sides are labelled by the six values of j in the symbol.

RELATIONS BETWEEN THE 6-j SYMBOLS AND THE V-C COEFFICIENTS. The orthogonal properties of the V-C coefficients may be used to obtain relations between 6-j symbols and the V-C coefficients starting from the definition (6.1.5). We have for example

(6.2.6)
$$(j_1 \ m_1 \ j_2 \ m_2 | j_1 \ j_2 \ j_{12} \ m_1+m_2)$$
$$\times (j_{12} \ m_1+m_2 \ j_3 \ m-m_1-m_2 | j_{12} \ j_3 \ j \ m)$$
$$= \sum_{j_{23}} (-1)^{j_1+j_2+j_3+j}[(2j_{12}+1)(2j_{23}+1)]^{\frac{1}{2}} \begin{Bmatrix} j_1 & j_2 & j_{12} \\ j_3 & j & j_{23} \end{Bmatrix}$$
$$\times (j_2 \ m_2 \ j_3 \ m-m_1-m_2 | j_2 \ j_3 \ j_{23} \ m-m_1)$$
$$\times (j_1 \ m_1 \ j_{23} \ m-m_1 | j_1 \ j_{23} \ j \ m)$$

(6.2.7)
$$(-1)^{j_1+j_2+j_3+j}[(2j_{12}+1)(2j_{23}+1)]^{\frac{1}{2}} \begin{Bmatrix} j_1 & j_2 & j_{12} \\ j_3 & j & j_{23} \end{Bmatrix}$$
$$\times (j_1 \ m_1 \ j_{23} \ m-m_1 | j_1 \ j_{23} \ j \ m)$$
$$= \sum_{m_2} (j_1 \ m_1 \ j_2 \ m_2 | j_1 \ j_2 \ j_{12} \ m_1+m_2)$$
$$\times (j_{12} \ m_1+m_2 \ j_3 \ m-m_1-m_2 | j_{12} \ j_3 \ j \ m)$$
$$\times (j_2 \ m_2 \ j_3 \ m-m_1-m_2 | j_2 \ j_3 \ j_{23} \ m-m_1)$$

We may also write equivalent and more symmetrical expressions with the 3-j symbols; for example,

(6.2.8)
$$\sum_{\mu_1 \mu_2 \mu_3} (-1)^{l_1+l_2+l_3+\mu_1+\mu_2+\mu_3} \begin{pmatrix} j_1 & l_2 & l_3 \\ m_1 & \mu_2 & -\mu_3 \end{pmatrix}$$
$$\times \begin{pmatrix} l_1 & j_2 & l_3 \\ -\mu_1 & m_2 & \mu_3 \end{pmatrix} \begin{pmatrix} l_1 & l_2 & j_3 \\ \mu_1 & -\mu_2 & m_3 \end{pmatrix}$$
$$= \begin{pmatrix} j_1 & j_2 & j_3 \\ m_1 & m_2 & m_3 \end{pmatrix} \begin{Bmatrix} j_1 & j_2 & j_3 \\ l_1 & l_2 & l_3 \end{Bmatrix}$$

6 · CONSTRUCTION OF INVARIANTS

ORTHOGONALITY AND SUM RULES. The known unitary nature of the recoupling transformations implies directly that the *real* 6-j symbols have the property

(6.2.9) $$\sum_{j} (2j+1)(2j''+1) \begin{Bmatrix} j_1 & j_2 & j' \\ j_3 & j_4 & j \end{Bmatrix} \begin{Bmatrix} j_3 & j_2 & j \\ j_1 & j_4 & j'' \end{Bmatrix} = \delta_{j'j''}$$

That is,

(6.2.10) $$[(2j+1)(2j'+1)]^{\frac{1}{2}} \begin{Bmatrix} j_1 & j_2 & j' \\ j_3 & j_4 & j \end{Bmatrix}$$

forms a real orthogonal matrix, rows and columns being labelled by j and j'.

Another relation for the 6-j symbols is given by composition of recoupling transformations. We have

$$\sum_{j_{23}} ((j_1\ j_2)j_{12},\ j_3,\ j|j_1,\ (j_2\ j_3)j_{23},\ j)(j_1,\ (j_2\ j_3)j_{23},\ j|j_2,\ (j_3\ j_1)j_{31},\ j)$$
$$= ((j_1\ j_2)j_{12},\ j_3,\ j|j_2,\ (j_3\ j_1)j_{31},\ j)$$

which yields

(6.2.11) $$\sum_{j_{23}} (-1)^{j_{23}+j_{31}+j_{12}}(2j_{23}+1) \begin{Bmatrix} j_1 & j_2 & j_{12} \\ j_3 & j & j_{23} \end{Bmatrix} \begin{Bmatrix} j_2 & j_3 & j_{23} \\ j_1 & j & j_{31} \end{Bmatrix}$$
$$= \begin{Bmatrix} j_3 & j_1 & j_{31} \\ j_2 & j & j_{12} \end{Bmatrix}$$

The sum rule of Biedenharn (1953) and Elliott (1953) is given by a similar consideration of the recoupling of four angular momenta. We take the transformation

$$((j_1\ j_2)j_{12},\ j_3,\ j_{123},\ j_4,\ j|(j_2\ j_3)j_{23},\ (j_1\ j_4)j_{14},\ j)$$

This is equal to the product of two successive recouplings of three angular momenta:

$$((j_1\ j_2)j_{12},\ j_3,\ j_{123},\ j_4,\ j|(j_2\ j_3)j_{23},\ j_1,\ j_{123},\ j_4,\ j)$$
$$\times ((j_2\ j_3)j_{23},\ j_1,\ j_{123},\ j_4,\ j|(j_2\ j_3)j_{23},\ (j_1\ j_4)j_{14},\ j)$$

We may alternatively carry out the recoupling in three stages, summing over the intermediate states containing j_{124}:

$$\sum_{j_{124}} ((j_1\ j_2)j_{12},\ j_3,\ j_{123},\ j_4,\ j|(j_1\ j_2)j_{12},\ j_4,\ j_{124},\ j_3,\ j)$$
$$\times ((j_1\ j_2)j_{12},\ j_4,\ j_{124},\ j_3,\ j|(j_1\ j_4)j_{14},\ j_2,\ j_{124},\ j_3,\ j)$$
$$\times ((j_1\ j_4)j_{14},\ j_2,\ j_{124},\ j_3,\ j|(j_2\ j_3)j_{23},\ (j_1\ j_4)j_{14},\ j)$$

Substitution of 6-j symbols into these two expressions gives

(6.2.12)
$$\begin{Bmatrix} j_1 & j_2 & j_{12} \\ j_3 & j_{123} & j_{23} \end{Bmatrix} \begin{Bmatrix} j_{23} & j_1 & j_{123} \\ j_4 & j & j_{14} \end{Bmatrix}$$
$$= \sum_{j_{124}} (-1)^{j_1+j_2+j_3+j_4+j_{12}+j_{23}+j_{14}+j_{123}+j+j_{124}}$$
$$\times (2j_{124}+1) \begin{Bmatrix} j_3 & j_2 & j_{23} \\ j_{14} & j & j_{124} \end{Bmatrix} \begin{Bmatrix} j_2 & j_1 & j_{12} \\ j_4 & j_{124} & j_{14} \end{Bmatrix} \begin{Bmatrix} j_3 & j_{12} & j_{123} \\ j_4 & j & j_{124} \end{Bmatrix}$$

a result which is used to obtain recursion relations for the 6-j symbols (see (6.3.5)).

OTHER NOTATIONS RELATED TO THE 6-j SYMBOL. The W coefficient of Racah (1942) is related to the 6-j symbol by

(6.2.13)
$$\begin{Bmatrix} j_1 & j_2 & j_3 \\ l_1 & l_2 & l_3 \end{Bmatrix} = (-1)^{j_1+j_2+l_1+l_2} W(j_1 \, j_2 \, l_2 \, l_1; j_3 \, l_3)$$

the U coefficient of Jahn (1951) by

(6.2.14)
$$\begin{Bmatrix} j_1 & j_2 & j_3 \\ l_1 & l_2 & l_3 \end{Bmatrix} = (-1)^{j_1+j_2+l_1+l_2} \frac{U(j_1 \, j_2 \, l_2 \, l_1; j_3 \, l_3)}{[(2j_3+1)(2l_3+1)]^{\frac{1}{2}}}$$

The choice of phase in the 6-j symbol has the advantage that the resulting quantity has symmetry properties which do not involve powers of -1 or other factors.

A related coefficient, used in angular distribution problems, is defined by Biedenharn, Blatt, and Rose (1952). It is

(6.2.15)
$$Z(a \, b \, c \, d; e \, f) = i^{f-a+c}[(2a+1)(2b+1)(2c+1)(2d+1)]^{\frac{1}{2}}$$
$$\times W(a \, b \, c \, d; e \, f)(a \, 0 \, c \, 0 | a \, c \, f \, 0)$$

(The quantity on the right is a V-C coefficient.)

6.3. Numerical Evaluation of the 6-j Symbol

FORMULAS FOR SPECIAL VALUES OF THE ARGUMENTS. Formulas in terms of the arguments are easily obtained from the defining relation (6.1.5) when one of the arguments is zero or $\frac{1}{2}$ or when one of the vector couplings involved has the form $j_1 + j_2 = J$.

We consider the case $l_1 + l_2 = j_3$, which includes the other two cases. We take $m_3 = j_3$ and $m_1 = -j_1$ or $m_2 = -j_2$. Then $m_{l_1} = -l_1$ and $m_{l_2} = l_2$ and the sum on the right reduces to one term; the 3-j symbols may be evaluated by the formula (3.7.11), giving finally

(3.1)
$$\begin{Bmatrix} j_1 & j_2 & l_1+l_2 \\ l_1 & l_2 & l_3 \end{Bmatrix} = (-1)^{j_1+j_2+l_1+l_2}$$
$$\times \left[\frac{(2l_1)!(2l_2)!(j_1+j_2+l_1+l_2+1)!(j_1+l_1+l_2-j_2)!(j_2+l_1+l_2-j_1)!(j_1+l_3-l_2)!(j_2+l_3-l_1)!}{(2l_1+2l_2+1)!(j_1+j_2-l_1-l_2)!(j_1+l_2-l_3)!(l_2+l_3-j_1)!(j_1+l_2+l_3+1)!(l_1+j_2-l_3)!(l_1+l_3-j_2)!(l_1+j_2+l_3+1)!} \right]^{\frac{1}{2}}$$

The special cases give

(6.3.2) $$\begin{Bmatrix} j_1 & j_2 & j_3 \\ 0 & j_3 & j_2 \end{Bmatrix} = (-1)^{j_1+j_2+j_3}[(2j_2+1)(2j_3+1)]^{-\frac{1}{2}}$$

(6.3.3) $$\begin{Bmatrix} j_1 & j_2 & j_3 \\ \frac{1}{2} & j_3-\frac{1}{2} & j_2+\frac{1}{2} \end{Bmatrix}$$
$$= (-1)^{j_1+j_2+j_3} \left[\frac{(j_1+j_3-j_2)(j_1+j_2-j_3+1)}{(2j_2+1)(2j_2+2)2j_3(2j_3+1)} \right]^{\frac{1}{2}}$$

(6.3.4) $$\begin{Bmatrix} j_1 & j_2 & j_3 \\ \frac{1}{2} & j_3-\frac{1}{2} & j_2-\frac{1}{2} \end{Bmatrix}$$
$$= (-1)^{j_1+j_2+j_3} \left[\frac{(j_1+j_2+j_3+1)(j_2+j_3-j_1)}{2j_2(2j_2+1)2j_3(2j_3+1)} \right]^{\frac{1}{2}}$$

Expressions for the 6-j symbol with other values of the arguments may be obtained from the above formulas by application of recursion relations derived from the sum rule (6.2.12).

RECURSION RELATIONS. We choose $\frac{1}{2}$ as one of the j values on the left of (6.2.12), which implies that the sum on the right reduces to two terms.

(6.3.5)
$$\begin{Bmatrix} a & b & c \\ d & e & f \end{Bmatrix} \begin{Bmatrix} b & f & d \\ \frac{1}{2} & d+\alpha & f+\beta \end{Bmatrix} (-1)^{\alpha+\beta-a-b-c}$$
$$= -(2e+1) \begin{Bmatrix} a & b & c \\ d+\alpha & e+\frac{1}{2} & f+\beta \end{Bmatrix}$$
$$\times \begin{Bmatrix} a & f & e \\ \frac{1}{2} & e+\frac{1}{2} & f+\beta \end{Bmatrix} \begin{Bmatrix} c & d & e \\ \frac{1}{2} & e+\frac{1}{2} & d+\alpha \end{Bmatrix}$$
$$+ 2e \begin{Bmatrix} a & b & c \\ d+\alpha & e-\frac{1}{2} & f+\beta \end{Bmatrix} \begin{Bmatrix} a & f & e \\ \frac{1}{2} & e-\frac{1}{2} & f+\beta \end{Bmatrix} \begin{Bmatrix} c & d & e \\ \frac{1}{2} & e-\frac{1}{2} & d+\alpha \end{Bmatrix}$$

where α and β take the values $\pm\frac{1}{2}$ independently. Substitution from the formulas (6.3.3) and (6.3.4) gives us a number of recursion relations.

We have, for example,

$$\begin{Bmatrix} a & b & c \\ d & e & f \end{Bmatrix}$$
$$\times [(a+b+c+1)(b+c-a)(c+d+e+1)(c+d-e)]^{\frac{1}{2}}$$

(6.3.6) $$= -2c[(b+d+f+1)(b+d-f)]^{\frac{1}{2}} \begin{Bmatrix} a & b-\frac{1}{2} & c-\frac{1}{2} \\ d-\frac{1}{2} & e & f \end{Bmatrix}$$

$$+ [(a+b-c+1)(a+c-b)(d+e-c+1)(c+e-d)]^{\frac{1}{2}}$$
$$\times \begin{Bmatrix} a & b & c-1 \\ d & e & f \end{Bmatrix}$$

TABULATION OF FORMULAS. Formulas derived in this way for 6-j symbols with smallest argument 1, $\frac{3}{2}$, or 2 are given in Table 5.

Similar tabulations for the W or U coefficients have been made, giving one argument the values $\frac{1}{2}$, 1, $\frac{3}{2}$, or 2 by Jahn (1951), Biedenharn et al. (1952), Biedenharn (1952) and Simon et al. (1954). The case of $\frac{5}{2}$ is given by Edmonds and Flowers (1952). A tabulation for $j = 3, \frac{7}{2}$, 4, $\frac{9}{2}$ has been published by Sato (1955).

GENERAL EXPRESSION FOR THE 6-j SYMBOL. A general formula has been obtained by Racah (1942). He introduced the series expression (3.6.11) for the V-C coefficient into the expression defining the invariant in terms of these quantities, and after a tedious calculation obtained a series with one index of summation. The result appears for the 6-j symbol as

(6.3.7)
$$\begin{Bmatrix} j_1 & j_2 & j_3 \\ l_1 & l_2 & l_3 \end{Bmatrix} = \Delta(j_1\ j_2\ j_3)\Delta(j_1\ l_2\ l_3)\Delta(l_1\ j_2\ l_3)\Delta(l_1\ l_2\ j_3) w \begin{Bmatrix} j_1 & j_2 & j_3 \\ l_1 & l_2 & l_3 \end{Bmatrix}$$

where

$$\Delta(a\ b\ c) = \left[\frac{(a+b-c)!(a-b+c)!(-a+b+c)!}{(a+b+c+1)!} \right]^{\frac{1}{2}}$$

and

$$\begin{Bmatrix} j_1 & j_2 & j_3 \\ l_1 & l_2 & l_3 \end{Bmatrix}$$
$$= \sum_z \frac{(-1)^z (z+1)!}{(z-j_1-j_2-j_3)!(z-j_1-l_2-l_3)!(z-l_1-j_2-l_3)!(z-l_1-l_2-j_3)! \times} $$
$$\times (j_1+j_2+l_1+l_2-z)!(j_2+j_3+l_2+l_3-z)!(j_3+j_1+l_3+l_1-z)!$$

and where the sum is over all positive integer values of z such that no factorial in the denominator has a negative argument. Schwinger (1952) has obtained a similar expression by another method.

NUMERICAL TABLES OF THE VALUES OF THE 6-j SYMBOL. There exist now extensive numerical tabulations of the W-coefficient. The tables of Biedenharn (1952) give the values exactly (i.e. as square roots of fractions). The ranges of values of arguments of his $W(l_1\ J_1\ l_2\ J_2;\ s\ L)$ are[3]

$$s : \tfrac{1}{2}(\tfrac{1}{2})3, \qquad L : 0(1)8; \qquad l_1, l_2 : 0(1)4; \qquad J_1, J_2 : \leq 4$$

Simon, Van der Sluis, and Biedenharn (1954) give the values to ten decimal places over a much wider range of arguments of $W(abcd;\ ef)$, namely

$$a : 0(\tfrac{1}{2})\tfrac{15}{2}; \qquad b : 0(\tfrac{1}{2})\tfrac{9}{2}; \qquad c : 0(\tfrac{1}{2})\tfrac{15}{2}$$

$$d : 0(\tfrac{1}{2})\tfrac{9}{2}; \qquad e : 0(\tfrac{1}{2})3; \qquad f : 0(1)8$$

Sharp et al. (1954) give $W(ljl'j';\ s\ k)$ in terms of prime factors of the numerator and denominator of the square of the coefficient:

$l, l' : 0(1)4;\ s : 0(\tfrac{1}{2})4;\ j = j';\ 0(\tfrac{1}{2})5$ and a few cases with $j \neq j'$ for $l, l', s = 0, 1, 2$. They give also $W(j\ j_1\ j\ j_1;\ L\ k)$ in prime factors for $j, j_1 = \tfrac{1}{2}(1)\tfrac{11}{2};\ L = 1, 2$.

The Z coefficient (6.2.15) which of course may be obtained easily from the above tables and the values of the 3-j symbol

$$\begin{pmatrix} j_1 & j_2 & j_3 \\ 0 & 0 & 0 \end{pmatrix}$$

(cf. (3.8)) has been tabulated for limited ranges of arguments by Biedenharn (1953) and by Sharp et al. (1954).

6.4. The 9-j Symbol

ANOTHER COUPLING SCHEME FOR FOUR ANGULAR MOMENTA. We have already considered one case of transformation between two coupling schemes of four angular momenta; this gave rise to the Elliott-Biedenharn sum rule (6.2.12) for 6-j symbols. The transformation we shall now deal with is of more general interest; the transformation coefficient may not in this case be expressed as a simple product of two transformations of type (6.1.3). The two types of state vectors to be considered which are built up from the four basic vectors $u(j_1\ m_1)$, $u(j_2\ m_2)$, $u(j_3\ m_3)$, and $u(j_4\ m_4)$ are $w((j_1\ j_2)j_{12}, (j_3\ j_4)j_{34},\ j\ m)$ and $w((j_1\ j_3)j_{13}, (j_2\ j_4)j_{24},\ j\ m)$. The transformation

$$((j_1\ j_2)j_{12}, (j_3\ j_4)j_{34},\ j\ m|(j_1\ j_3)j_{13}, (j_2\ j_4)j_{24},\ j\ m)$$

which connects these two schemes may be performed in three steps; a dummy index j' is involved which we sum over in the final expression. In each of the steps recoupling of only three angular momenta is carried out, so that the transformation coefficient may be expressed in terms of the 6-j symbols. We have thus, dropping as usual the superfluous magnetic quantum numbers,

---

[3] Using the usual convention where the number in brackets is the interval of tabulation; i.e. $\tfrac{1}{2}(\tfrac{1}{2})3$ implies the values $\tfrac{1}{2}, 1, \tfrac{3}{2}, 2. \tfrac{5}{2}, 3$ and $0(1)4$ implies 0, 1, 2, 3, 4. See also Howell (1959), Rotenberg et al. (1959).

6.4 · THE 9-j SYMBOL

$$((j_1\ j_2)j_{12},\ (j_3\ j_4)j_{34},\ j|(j_1\ j_3)j_{13},\ (j_2\ j_4)j_{24},\ j)$$

$$= \sum_{j'} ((j_1\ j_2)j_{12},\ j_{34},\ j|j_1,\ (j_2\ j_{34})j',\ j)$$

$$\times (j_2,\ (j_3\ j_4)j_{34},\ j'|j_3,\ (j_2\ j_4)j_{24},\ j')$$

(6.4.1)
$$\times (j_1,\ (j_3\ j_{24})j',\ j|(j_1\ j_3)j_{13},\ j_{24},\ j)$$

$$= [(2j_{12} + 1)(2j_{34} + 1)(2j_{13} + 1)(2j_{24} + 1)]^{\frac{1}{2}}$$

$$\times \sum_{j'} (-1)^{2j'}(2j' + 1)$$

$$\times \begin{Bmatrix} j_1 & j_2 & j_{12} \\ j_{34} & j & j' \end{Bmatrix} \begin{Bmatrix} j_3 & j_4 & j_{34} \\ j_2 & j' & j_{24} \end{Bmatrix} \begin{Bmatrix} j_{13} & j_{24} & j \\ j' & j_1 & j_3 \end{Bmatrix}$$

Such a recoupling scheme occurs quite frequently, and we are led to the examination of yet another kind of rotational invariant.

DEFINITION OF THE 9-j SYMBOL. The 9-j symbol[4] is defined by the relation

(6.4.2)
$$((j_1\ j_2)j_{12},\ (j_3\ j_4)j_{34},\ j|(j_1\ j_3)j_{13},\ (j_2\ j_4)j_{24},\ j)$$

$$= [(2j_{12} + 1)(2j_{34} + 1)(2j_{13} + 1)(2j_{24} + 1)]^{\frac{1}{2}} \begin{Bmatrix} j_1 & j_2 & j_{12} \\ j_3 & j_4 & j_{34} \\ j_{13} & j_{24} & j \end{Bmatrix}$$

I.e. we have, choosing a symmetrical set of labels for the j's:

(6.4.3)
$$\begin{Bmatrix} j_{11} & j_{12} & j_{13} \\ j_{21} & j_{22} & j_{23} \\ j_{31} & j_{32} & j_{33} \end{Bmatrix} = \sum_{\kappa} (-1)^{2\kappa}(2\kappa + 1)$$

$$\times \begin{Bmatrix} j_{11} & j_{21} & j_{31} \\ j_{32} & j_{33} & \kappa \end{Bmatrix} \begin{Bmatrix} j_{12} & j_{22} & j_{32} \\ j_{21} & \kappa & j_{23} \end{Bmatrix} \begin{Bmatrix} j_{13} & j_{23} & j_{33} \\ \kappa & j_{11} & j_{12} \end{Bmatrix}$$

If we replace the 6-j symbols by the appropriate 3-j symbols, making use of their orthogonality properties, we find the remarkably symmetric expression

(6.4.4)
$$\begin{Bmatrix} j_{11} & j_{12} & j_{13} \\ j_{21} & j_{22} & j_{23} \\ j_{31} & j_{32} & j_{33} \end{Bmatrix}$$

$$= \sum_{\text{all } m\text{'s}} \begin{pmatrix} j_{11} & j_{12} & j_{13} \\ m_{11} & m_{12} & m_{13} \end{pmatrix} \begin{pmatrix} j_{21} & j_{22} & j_{23} \\ m_{21} & m_{22} & m_{23} \end{pmatrix} \begin{pmatrix} j_{31} & j_{32} & j_{33} \\ m_{31} & m_{32} & m_{33} \end{pmatrix}$$

$$\times \begin{pmatrix} j_{11} & j_{21} & j_{31} \\ m_{11} & m_{21} & m_{31} \end{pmatrix} \begin{pmatrix} j_{12} & j_{22} & j_{32} \\ m_{12} & m_{22} & m_{32} \end{pmatrix} \begin{pmatrix} j_{13} & j_{23} & j_{33} \\ m_{13} & m_{23} & m_{33} \end{pmatrix}$$

---

[4] Wigner (1951). For numerical tabulations see Smith (1958) and Howell (1959).

We can cast some light on the significance of this 9-j symbol by returning to the discussion in (6.1) of the contraction process on products of 3-j symbols. The next most complicated contraction diagram after that associated with the 6-j symbol would appear to be that in Fig. 6.3.

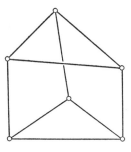

Fig. 6.3

However this diagram may be shown to correspond to the product of two 6-j symbols appearing in the Biedenharn-Elliott sum rule. Another diagram with 9-j values which satisfies the conditions (3 lines leave each vertex, each line terminates at two vertices) is the linkage of Fig. 6.4.

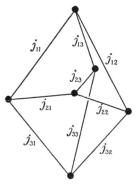

Fig. 6.4

Reference to the expression (6.4.4) for the 9-j symbol in terms of 3-j symbols shows that this diagram does indeed correspond to the 9-j symbol; the labels on the lines give a possible assignment of j values.

THE SYMMETRIES OF THE 9-j SYMBOL. It is clear that we may permute the rows or columns in the matrix forming the 9-j symbol, or transpose the matrix itself, producing at most a change of sign of the numerical value. An *odd* permutation of the rows produces an odd permutation of the j's in each of the last three of the 3-j symbols in

(6.4.4), with no change in the first three; an odd permutation of the columns, on the other hand, gives an odd permutation of the j's in each of the first three 3-j symbols. Transposition (i.e. replacing rows by columns and vice versa) merely alters the ordering of the 3-j symbols in the product. Hence the symmetry properties of the 3-j symbols (3.7.4) and (3.7.5) show us that an odd permutation of rows or columns produces a sign change of

(6.4.5) $\qquad (-1)^{j_{11}+j_{12}+j_{13}+j_{21}+j_{22}+j_{23}+j_{31}+j_{32}+j_{33}}$

An even permutation or a transposition clearly leaves the symbol unchanged. The symmetry group[5] may easily be shown to have 72 elements, being the product[6] of the three permutation groups of three, three, and two objects respectively; i.e. $G = S_3 \times S_3 \times S_2$.

ORTHOGONALITY AND SUM RULES. These are derived in exactly the same way as for the 6-j symbol; we have

(6.4.6)
$$\sum_{j_{12}j_{34}} (2j_{12} + 1)(2j_{34} + 1)(2j_{13} + 1)(2j_{24} + 1)$$
$$\begin{Bmatrix} j_1 & j_2 & j_{12} \\ j_3 & j_4 & j_{34} \\ j_{13} & j_{24} & j \end{Bmatrix} \begin{Bmatrix} j_1 & j_2 & j_{12} \\ j_3 & j_4 & j_{34} \\ j'_{13} & j'_{24} & j \end{Bmatrix} = \delta_{j_{13}j'_{13}}\delta_{j_{24}j'_{24}},$$

from the unitary property of the recoupling transformation on four angular momenta. The multiplicative property of the transformations:

$$\sum_{j_{12}j_{34}} ((j_1\ j_2)j_{12}, (j_3\ j_4)j_{34}, j|(j_1\ j_3)j_{13}, (j_2\ j_4)j_{24}, j)$$
$$\times ((j_1\ j_3)j_{13}, (j_2\ j_4)j_{24}, j|(j_1\ j_4)j_{14}, (j_2\ j_3)j_{23}, j)$$
$$= ((j_1\ j_2)j_{12}, (j_3\ j_4)j_{34}, j|(j_1\ j_4)j_{14}, (j_2\ j_3)j_{23}, j)$$

gives the sum rule

(6.4.7)
$$\sum_{j_{13}j_{24}} (-1)^{2j_{13}+j_{24}+j_{23}-j_{34}}$$
$$\times \begin{Bmatrix} j_1 & j_2 & j_{12} \\ j_3 & j_4 & j_{34} \\ j_{13} & j_{24} & j \end{Bmatrix} \begin{Bmatrix} j_1 & j_3 & j_{13} \\ j_4 & j_2 & j_{24} \\ j_{14} & j_{23} & j \end{Bmatrix} (2j_{13} + 1)(2j_{24} + 1)$$
$$= \begin{Bmatrix} j_1 & j_2 & j_{12} \\ j_4 & j_3 & j_{34} \\ j_{14} & j_{23} & j \end{Bmatrix}$$

---

[5]Cf. Jahn and Hope (1954).
[6]See Littlewood (1950).

Relations linking the 6-j symbols and the 9-j symbols may be obtained by making use of the orthogonality properties of the 6-j symbols or by the composition of recoupling transformations. We get for example from the expression (6.4.3) by use of the orthogonality of the 6-j symbols (6.2.9) a relation which is of some use in computing numerical values of 9-j symbols.

(6.4.8)
$$\sum_\mu (2\mu + 1) \begin{Bmatrix} j_{11} & j_{12} & \mu \\ j_{21} & j_{22} & j_{23} \\ j_{31} & j_{32} & j_{33} \end{Bmatrix} \begin{Bmatrix} j_{11} & j_{12} & \mu \\ j_{23} & j_{33} & \lambda \end{Bmatrix}$$
$$= (-1)^{2\lambda} \begin{Bmatrix} j_{21} & j_{22} & j_{23} \\ j_{12} & \lambda & j_{32} \end{Bmatrix} \begin{Bmatrix} j_{31} & j_{32} & j_{33} \\ \lambda & j_{11} & j_{21} \end{Bmatrix}$$

FURTHER REMARKS ON THE RECOUPLING OF FOUR ANGULAR MOMENTA. We have already seen how two different recoupling coefficients associated with four angular momenta have been given in terms of a product of two 6-j symbols and a sum of products of three 6-j symbols (the 9-j symbol) respectively. It may be shown that every coefficient associated with a recoupling of four angular momenta which is not simply a recoupling of three of the four may be expressed in one or other of the above ways. For example we have

(6.4.9)
$$((j_1\ j_2)j_{12},\ j_3,\ j_{123},\ j_4,\ j|(j_4\ j_2)j_{24},\ j_3,\ j_{234},\ j_1,\ j)$$
$$= (-1)^{j_1+j_4+j_{123}+j_{234}+2j}$$
$$\times [(2j_{12}+1)(2j_{123}+1)(2j_{24}+1)(2j_{234}+1)]^{\frac{1}{2}}$$
$$\times \begin{Bmatrix} j_2 & j_4 & j_{24} \\ j_1 & j & j_{234} \\ j_{12} & j_{123} & j_3 \end{Bmatrix}$$

The recoupling of four angular momenta is evidently involved in the transition from LS to jj coupling; see for example Condon and Shortley (1935) and, for the application of the 9-j symbol, Edmonds and Flowers (1952). It arises also in the evaluation of matrix elements of tensor products of tensor operators, a subject dealt with in the next chapter. The recoupling coefficients are also very important in the computation of fractional parentage coefficients; these computations sometimes involve the recoupling of *five* angular momenta, which brings in the 12-j *symbols*, to be mentioned shortly. The reader is referred to the papers of Elliott (1953) and Jahn (1954).

OTHER NOTATIONS FOR THE 9-j SYMBOL. The χ function of Hope and Jahn (cf. Hope (1951), Jahn and Hope (1954)) is defined directly in terms of the recoupling coefficient. We have thus

$$
(6.4.10) \quad [(2e+1)(2f+1)(2g+1)(2h+1)]^{\frac{1}{2}} \begin{Bmatrix} a & b & e \\ c & d & f \\ g & h & k \end{Bmatrix}
$$
$$= \chi(a\ b\ c\ d;e\ f;g\ h;k)$$

The S function of Schwinger (1952) is given by

$$
(6.4.11) \quad (-1)^{j_1+j_4-j_{12}-j_{24}} \begin{Bmatrix} j_1 & j_2 & j_{12} \\ j_3 & j_4 & j_{34} \\ j_{13} & j_{24} & j \end{Bmatrix}
$$
$$= S(j_1\ j_2\ j_3\ j_4;j_{12}\ j_{34}\ j_{13}\ j_{24};j)$$

Fano's X function is identical with the 9-j symbol, and is written alternatively (cf. Fano (1952))

$$
(6.4.12) \quad X \begin{Bmatrix} j_1 & j_2 & j \\ j_3 & j_4 & j' \\ k & k' & J \end{Bmatrix} \equiv X(j_1\ j_2\ j;j_3\ j_4\ j';k\ k'\ J)
$$

The coefficient of Coester and Jauch (1953) is the same as the recoupling coefficient (6.4.1). Hence we have

$$(c\ c'|\Gamma(a\ a'\ b\ b'\ d)|e\ f) \equiv ((a\ b)c, (a'\ b')c', d|(a\ a')e, (b\ b')f, d)$$
$$
(6.4.13) \quad = [(2c+1)(2c'+1)(2e+1)(2f+1)]^{\frac{1}{2}} \begin{Bmatrix} a & b & c \\ a' & b' & c' \\ e & f & d \end{Bmatrix}
$$

EVALUATION OF THE 9-j SYMBOL. The expression (6.4.3) giving the 9-j symbol in terms of 6-j symbols shows that a 9-j symbol with one argument zero reduces to a 6-j symbol times a factor:

$$
\begin{Bmatrix} a & b & e \\ c & d & e \\ f & f & 0 \end{Bmatrix} = \begin{Bmatrix} 0 & e & e \\ f & d & b \\ f & c & a \end{Bmatrix} = \begin{Bmatrix} e & 0 & e \\ c & f & a \\ d & f & b \end{Bmatrix}
$$
$$
(6.4.14) \qquad = \begin{Bmatrix} f & f & 0 \\ d & c & e \\ b & a & e \end{Bmatrix} = \begin{Bmatrix} f & b & d \\ 0 & e & e \\ f & a & c \end{Bmatrix} = \begin{Bmatrix} a & f & c \\ e & 0 & e \\ b & f & d \end{Bmatrix}
$$

$$= \begin{Bmatrix} b & a & e \\ f & f & 0 \\ d & c & e \end{Bmatrix} = \begin{Bmatrix} e & d & c \\ e & b & a \\ 0 & f & f \end{Bmatrix} = \begin{Bmatrix} c & e & d \\ a & e & b \\ f & 0 & f \end{Bmatrix}$$

$$= \frac{(-1)^{b+c+e+f}}{[(2e+1)(2f+1)]^{\frac{1}{2}}} \begin{Bmatrix} a & b & e \\ d & c & f \end{Bmatrix}$$

where the symmetry properties of the 9-j symbol have been used to cover all cases.

Evaluation of the 6-j symbols in (6.4.3) is one way of determining 9-j symbols with no zero arguments. The symbol should be arranged so that the smallest argument $j_{\min}$ does not fall in the positions 13, 22 or 31. If this is done, the sum over κ has at most $(2j_{\min} + 1)$ terms, and if $j_{\min} \leq 2$ we may make use of the formulas of Table 5, to give a fairly simple expression in terms of 6-j symbols. For example we get for the general 9-j symbol with $j_{\min} = \frac{1}{2}$:

(6.4.15)
$$\begin{Bmatrix} \frac{1}{2} & b & b+\frac{1}{2} \\ d & e & f \\ d+\frac{1}{2} & h & k \end{Bmatrix}$$

$$= \frac{(-1)^{b+d+f+h}}{(2k+1)[(2b+1)(2b+2)(2d+1)(2d+2)]^{\frac{1}{2}}}$$

$$\times \begin{bmatrix} [(-b+f+k+\frac{1}{2})(b+f-k+\frac{1}{2})(d+h-k+\frac{1}{2}) \\ \qquad \times (-d+h+k+\frac{1}{2})]^{\frac{1}{2}} \begin{Bmatrix} d & e & f \\ b & k+\frac{1}{2} & h \end{Bmatrix} \\ + [(b+f+k+\frac{3}{2})(b-f+k+\frac{1}{2})(d+h+k+\frac{3}{2}) \\ \qquad \times (d-h+k+\frac{1}{2})]^{\frac{1}{2}} \begin{Bmatrix} d & e & f \\ b & k-\frac{1}{2} & h \end{Bmatrix} \end{bmatrix}$$

However (6.4.8) sometimes gives a simpler expression in the 6-j symbols; see for example (6.4.17).

Numerical values of certain 9-j symbols have been tabulated by Sharp et al. (1954)

$$\begin{Bmatrix} a & b & c \\ a' & b & c \\ g & h & k \end{Bmatrix}$$ has been given for $a, a' = 1, 2$; $b, c : 1(\frac{1}{2})5$ and g, h, k with even integer values less than 9.

These choices of arguments are used in analysis of triple correlations of nuclear radiations in which the intermediate radiation is a gamma ray.

$\begin{Bmatrix} a & b & c \\ a & e & f \\ k & k & 1 \end{Bmatrix}$ has been given for reactions with polarized particles with channel spins $\leq \frac{5}{2}$ and orbital angular momenta ≤ 3.

THE 12-j SYMBOLS. In the theory of fractional parentage coefficients it is sometimes necessary to consider recoupling of five angular momenta, and the 12-j symbols have therefore been introduced. There are two distinct types of symbol, corresponding to the respective diagrams I and II in Fig. 6.5. The properties and applications of these quantities

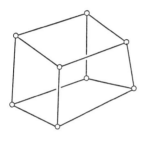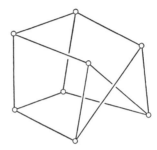

Fig. 6.5

are discussed in the papers of Jahn and Hope (1954), Ord-Smith (1954) and Elliott and Flowers (1955).

COMPUTATION OF LS-jj COUPLING COEFFICIENTS.[7] The LS-jj coupling transformation coefficient is given by

$$((l_1 \; l_2)L, (s_1 \; s_2)S, J|(l_1 \; s_1)j_1, (l_2 \; s_2)j_2, J)$$

(6.4.16)
$$= [(2L + 1)(2S + 1)(2j_1 + 1)(2j_2 + 1)]^{\frac{1}{2}} \begin{Bmatrix} l_1 & l_2 & L \\ s_1 & s_2 & S \\ j_1 & j_2 & J \end{Bmatrix}$$

Since $s_1 = s_2 = \frac{1}{2}$, S may take the values 0 or 1. If $S = 0$ the right-hand side reduces easily by use of (6.4.14) to

$$(-1)^{l_1+j_2+J+\frac{1}{2}} \left[\frac{(2j_1 + 1)(2j_2 + 1)}{2} \right]^{\frac{1}{2}} \begin{Bmatrix} J & j_1 & j_2 \\ \frac{1}{2} & l_2 & l_1 \end{Bmatrix}$$

We may evaluate the 9-j symbol for $S = 1$ by use of (6.4.8) and (6.4.14):

---

[7] Cf. Condon and Shortley (1935).

(6.4.17)
$$3\begin{Bmatrix} J & L & 1 \\ \frac{1}{2} & \frac{1}{2} & \lambda \end{Bmatrix} \begin{Bmatrix} l_1 & l_2 & L \\ \frac{1}{2} & \frac{1}{2} & 1 \\ j_1 & j_2 & J \end{Bmatrix} + \frac{(-1)^{l_1+j_2-\lambda}}{2(2J+1)} \begin{Bmatrix} J & l_2 & l_1 \\ \frac{1}{2} & j_1 & j_2 \end{Bmatrix} \delta_{LJ}$$
$$= (-1)^{2\lambda} \begin{Bmatrix} j_1 & j_2 & J \\ \frac{1}{2} & \lambda & l_2 \end{Bmatrix} \begin{Bmatrix} l_2 & l_1 & L \\ \frac{1}{2} & \lambda & j_1 \end{Bmatrix}$$

where λ is given the value $(L+J)/2$ if $L \neq J$ or $L+\frac{1}{2}$ if $L=J$. We get for example the transformation coefficient

$$((l_1\ l_2)L, (\tfrac{1}{2}\ \tfrac{1}{2})1, J = L+1 | (l_1\ \tfrac{1}{2})l_1+\tfrac{1}{2}, (l_2\ \tfrac{1}{2})l_2+\tfrac{1}{2}, J = L+1)$$
$$= \left[\frac{(l_1+l_2+J+1)(l_1+l_2+J+2)(l_1-l_2+J)(-l_1+l_2+J)}{2J(2J+1)(2l_1+1)(2l_2+1)} \right]^{\frac{1}{2}}$$

A number of new relations between 6-j and 9-j symbols are given in the paper of Innes and Ufford (1958). One of these relations effectively defines the 15-j symbol corresponding to the striking diagram in Fig. 6.6.

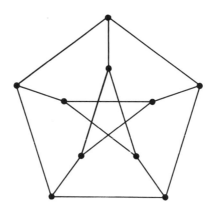

Fig. 6.6.

CHAPTER 7

The Evaluation of Matrix Elements in Actual Problems

7.1. Matrix Elements of the Tensor Product of Two Tensor Operators

TENSOR OPERATORS OPERATING ON THE SAME SYSTEM. The tensor operators $\mathsf{T}(k_1)$ and $\mathsf{T}(k_2)$ are built up from the same coordinates, momenta, etc.

The reduced matrix element of the tensor product $\mathsf{X}(K)$ of $\mathsf{T}(k_1)$ and $\mathsf{T}(k_2)$ is given by

$$(\gamma'\ j'||\mathsf{X}(K)||\gamma\ j) = \sum_{q_1 q_2 Q m} (k_1\ q_1\ k_2\ q_2|k_1\ k_2\ K\ Q)(K\ Q\ j\ m|K\ j\ j'\ m')$$
$$\times (-1)^{K-j+j'}(2j'+1)^{\frac{1}{2}} \sum_{\gamma''j''m''} (\gamma'\ j'\ m'|T(k_1\ q_1)|\gamma''\ j''\ m'')$$
$$\times (\gamma''\ j''\ m''|T(k_2\ q_2)|\gamma\ j\ m)$$

The product of the reduced matrix elements of the individual operators is on the other hand

$$(\gamma'\ j'||\mathsf{T}(k_1)||\gamma''\ j'')(\gamma''\ j''||\mathsf{T}(k_2)||\gamma\ j)$$
$$= \sum_{q_1 q_2 m m''} (k_1\ q_1\ j''\ m''|k_1\ j''\ j'\ m')(k_2\ q_2\ j\ m|k_2\ j\ j''\ m'')$$
$$\times (-1)^{k_1+k_2-j+j'}[(2j'+1)(2j''+1)]^{\frac{1}{2}}$$
$$\times (\gamma'\ j'\ m'|T(k_1\ q_1)|\gamma''\ j''\ m'')(\gamma''\ j''\ m''|T(k_2\ q_2)|\gamma\ j\ m)$$

We have made use here of (5.4.2) and the symmetry properties (3.5) of the V-C coefficients. Attention should be paid to the indices over which summations are carried out.

On inspection of the two above equations we see that two coupling schemes of the three angular momenta k_1, k_2, and j are involved. Thus we may apply (6.1.5) and relate the two reduced matrix element expressions by means of the 6-j symbol

(7.1.1)
$$(\gamma'\ j'||\mathsf{X}(K)||\gamma\ j) = (2K+1)^{\frac{1}{2}}(-1)^{K+j+j'} \sum_{\gamma''j''} \begin{Bmatrix} k_1 & k_2 & K \\ j & j' & j'' \end{Bmatrix}$$
$$\times (\gamma'\ j'||\mathsf{T}(k_1)||\gamma''\ j'')(\gamma''\ j''||\mathsf{T}(k_2)||\gamma\ j)$$

TENSOR OPERATORS OPERATING ON DIFFERENT SYSTEMS. The tensor operators $\mathsf{T}(k_1)$ and $\mathsf{U}(k_2)$ are supposed to work on parts 1 and 2 respectively of a system, i.e. they commute, and we shall derive an expression for the reduced matrix element of the tensor product in the coupled scheme in terms of the reduced matrix elements of the individual operators in the uncoupled scheme. The quantum numbers $j_1\, m_1$, $j_2\, m_2$, and $J\, M$ refer to the parts 1 and 2 and the whole system respectively.

We first make use of (5.4.2) and (3.5.14) to obtain an expression for the reduced matrix element of $\mathbf{X}(K)$ in the coupled scheme $(\gamma\, j_1\, j_2\, J\, M)$ in terms of the matrix element in the uncoupled scheme $(\gamma\, j_1\, m_1\, j_2\, m_2)$:

$$(\gamma'\, j_1'\, j_2'\, J'||\mathbf{X}(K)||\gamma\, j_1\, j_2\, J)(2K+1)^{-\frac{1}{2}}$$

(7.1.2)
$$= \sum_{QMM'm_1m_2m_1'm_2'} (J'\, M'\, J\, M|J'\, J\, K\, Q)(j_1\, m_1\, j_2\, m_2|j_1\, j_2\, J\, M)$$
$$\times (j_1'\, m_1'\, j_2'\, m_2'|j_1'\, j_2'\, J'\, M')(-1)^{j_1+j_2+m_1+m_2}$$
$$\times (\gamma'\, j_1'\, m_1'\, j_2'\, m_2'|X(K\, Q)|\gamma\, j_1\, -m_1\, j_2\, -m_2)$$

The reduced matrix element of the simple product of $T(k_1\, q_1)$ and $U(k_2\, q_2)$ is now expressed in terms of the same matrix element of $X(K\, Q)$; we use here (5.4.2) and (5.1.9).

$$(\gamma'\, j_1'\, j_2'||\mathbf{T}(k_1)\mathbf{U}(k_2)||\gamma\, j_1\, j_2)[(2k_1+1)(2k_2+1)]^{-\frac{1}{2}}$$
$$= \sum_{\gamma''} (\gamma'\, j_1'\, j_2'||\mathbf{T}(k_1)||\gamma''\, j_1\, j_2')$$

(7.1.3)
$$\times (\gamma''\, j_1\, j_2'||\mathbf{U}(k_2)||\gamma\, j_1\, j_2)[(2k_1+1)(2k_2+1)]^{-\frac{1}{2}}$$
$$= \sum_{m_1m_1'm_2m_2'q_1q_2Q} (j_1'\, m_1'\, j_1\, m_1|j_1'\, j_1\, k_1\, q_1)(j_2'\, m_2'\, j_2\, m_2|j_2'\, j_2\, k_2\, q_2)$$
$$\times (k_1\, q_1\, k_2\, q_2|k_1\, k_2\, K\, Q)\cdot(-1)^{j_1+j_2+m_1+m_2}$$
$$\times (\gamma'\, j_1'\, m_1'\, j_2'\, m_2'|X(K\, Q)|\gamma\, j_1\, -m_1\, j_2\, -m_2)$$

We see immediately that these two expressions are associated with different coupling schemes for the angular momenta $j_1'\, j_1$, j_2', j_2 and the left-hand sides are related by the corresponding transformation coefficient

$$(\gamma'\, j_1'\, j_2'\, J'||\mathbf{X}(K)||\gamma\, j_1\, j_2\, J)(2K+1)^{-\frac{1}{2}}$$
$$= \sum_{\gamma''} (\gamma'\, j_1'||\mathbf{T}(k_1)||\gamma''\, j_1)$$

(7.1.4)
$$\times (\gamma''\, j_2'||\mathbf{U}(k_2)||\gamma\, j_2)[(2k_1+1)(2k_2+1)]^{-\frac{1}{2}}$$
$$\times ((j_1'\, j_1)k_1,\, (j_2'\, j_2)k_2,\, K|(j_1'\, j_2')J',\, (j_1\, j_2)J,\, K)$$

Thus (6.4.2) gives the desired relation involving the 9-j symbol:

$$(\gamma'\, j_1'\, j_2'\, J'||\mathbf{X}(K)||\gamma\, j_1\, j_2\, J)$$
$$= \sum_{\gamma''} (\gamma'\, j_1'||\mathbf{T}(k_1)||\gamma''\, j_1)(\gamma''\, j_2'||\mathbf{U}(k_2)||\gamma\, j_2)$$

(7.1.5)
$$\times [(2J+1)(2J'+1)(2K+1)]^{\frac{1}{2}} \begin{Bmatrix} j_1' & j_1 & k_1 \\ j_2' & j_2 & k_2 \\ J' & J & K \end{Bmatrix}$$

7.1 · MATRIX ELEMENTS

A number of useful relations may now be obtained by specializing this formula, making use of the expressions (6.4.14) for the 9-j symbols with one argument zero.

SCALAR PRODUCT OF TWO COMMUTING TENSOR OPERATORS. The matrix element of the scalar product $(\mathsf{T}(k) \cdot \mathsf{U}(k))$ in the scheme $(\gamma j_1 j_2 J M)$ is gotten by setting $K = 0$ and $k_1 = k_2 = k$ in (7.1.5).

$$(\gamma'\ j_1'\ j_2'\ J'\ M'|(\mathsf{T}(k)\cdot\mathsf{U}(k))|\gamma\ j_1\ j_2\ J\ M)$$

(7.1.6)
$$= (-1)^{j_1+j_2'+J} \delta_{J'J} \delta_{M'M} \begin{Bmatrix} J & j_2' & j_1' \\ k & j_1 & j_2 \end{Bmatrix}$$

$$\times \sum_{\gamma''} (\gamma'\ j_1'||\mathsf{T}(k)||\gamma''\ j_1)(\gamma''\ j_2'||\mathsf{U}(k)||\gamma\ j_2)$$

SINGLE OPERATOR IN COUPLED SCHEME. We obtain the reduced matrix element of a tensor operator $\mathsf{T}(k)$ working only on part 1 in the coupled scheme $(\gamma\ j_1\ j_2\ J\ M)$. We put $k_2 = 0$ in (7.1.5), substituting $\mathsf{U}(k) = 1$.

$$(\gamma'\ j_1'\ j_2\ J'||\mathsf{T}(k)||\gamma\ j_1\ j_2\ J)$$

(7.1.7)
$$= (-1)^{j_1'+j_2+J+k}[(2J+1)(2J'+1)]^{\frac{1}{2}} \begin{Bmatrix} j_1' & J' & j_2 \\ J & j_1 & k \end{Bmatrix}$$

$$\times (\gamma'\ j_1'||\mathsf{T}(k)||\gamma\ j_1)$$

In the same way for a tensor operator $\mathsf{U}(k)$ working only on part 2,

$$(\gamma'\ j_1\ j_2'\ J'||\mathsf{U}(k)||\gamma\ j_1\ j_2\ J)$$

(7.1.8)
$$= (-1)^{j_1+j_2+J'+k}[(2J+1)(2J'+1)]^{\frac{1}{2}} \begin{Bmatrix} j_2' & J' & j_1 \\ J & j_2 & k \end{Bmatrix}$$

$$\times (\gamma'\ j_2'||\mathsf{U}(k)||\gamma\ j_2)$$

MATRIX ELEMENTS OF ANGULAR MOMENTUM L_1 IN SCHEME $(l_1\ l_1\ l\ m)$. We take a simple example of the application of (7.1.7); we compute the expectation value of the z component of $\mathbf{L}_1$ in the scheme defined by the vector addition $\mathbf{L} = \mathbf{L}_1 + \mathbf{L}_2$.

We have for the reduced matrix element

$(l_1\ l_2\ l||\mathbf{L}_1||l_1\ l_2\ l)$

$$= \hbar(-1)^{l_1+l_2+l+1}(2l+1) \begin{Bmatrix} l_1 & l & l_2 \\ l & l_1 & 1 \end{Bmatrix} [(2l_1+1)(l_1+1)l_1]^{\frac{1}{2}}$$

by use of (5.4.3). Evaluation of the 6-j symbol gives

$$\frac{(l_1\ l_2\ l||\mathbf{L}_1||l_1\ l_2\ l)}{(2l+1)^{\frac{1}{2}}} = \frac{l_1(l_1+1) + l(l+1) - l_2(l_2+1)}{2[l(l+1)]^{\frac{1}{2}}}$$

7 · EVALUATION OF MATRIX ELEMENTS

This corresponds to the projection of $\mathbf{L}_1$ onto the $\mathbf{L}$ axis, given in the semiclassical procedure by

$$\frac{(\mathbf{L}_1 \cdot \mathbf{L})}{[l(l+1)]^{\frac{1}{2}}} = \frac{\mathbf{L}_1^2 + \mathbf{L}^2 - \mathbf{L}_2^2}{2[l(l+1)]^{\frac{1}{2}}} = \frac{l_1(l_1+1) + l(l+1) - l_2(l_2+1)}{2[l(l+1)]^{\frac{1}{2}}}$$

The expectation value of L_{1z} in the state $l_1 \, l_2 \, l \, m$ is given by (5.4.1)

$$(l_1 \, l_2 \, l \, m | L_{1z} | l_1 \, l_2 \, l \, m) = (-1)^{l-m} \begin{pmatrix} l & 1 & l \\ -m & 0 & m \end{pmatrix} (l_1 \, l_2 \, l \, ||L_1|| \, l_1 \, l_2 \, l)$$

$$= \frac{m\{l_1(l_1+1) + l(l+1) - l_2(l_2+1)\}}{2l(l+1)}$$

The result corresponds to the classical idea that $\mathbf{L}_1$ is in a state of precession about the direction of $\mathbf{L}$, and that the mean value $\overline{L_{1z}}$ of the projection of $\mathbf{L}_1$ onto the z axis is obtained by first projecting $\mathbf{L}_1$ onto $\mathbf{L}$ and then onto the z axis. (See Fig. 7.1)

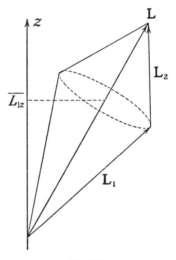

Fig. 7.1

ZEEMAN EFFECT. The above results may be used in computing the Zeeman splitting of atomic spectra with a weak field (cf. Condon and Shortley (1935) p. 150). If the field is along the z axis we have in Condon and Shortley's notation,

$$H^M = \frac{e\mathcal{IC}}{2\mu c}(L_z + 2S_z)$$

This quantity is to be computed for a state with definite $\mathbf{J} = \mathbf{L} + \mathbf{S}$, and we get

$$\frac{2\mu c}{e\mathcal{K}} (S L J M|H^M|S L J M) = \begin{pmatrix} J & 1 & J \\ -M & 0 & M \end{pmatrix} [(S L J||\mathbf{L}||S L J)$$
$$+ 2(S L J||\mathbf{S}||S L J)]$$
$$= \frac{M}{2J(J+1)} [3J(J+1) - L(L+1) + S(S+1)]$$

The reader may show for himself that Condon and Shortley's result on p. 151 is gotten by setting $S = \frac{1}{2}$, $L = l$, and $J = l \pm \frac{1}{2}$.

MATRIX ELEMENTS OF THE SPHERICAL HARMONICS IN jj-COUPLING.[1] The reduced matrix elements of $\mathbf{C}^{(k)}$ (for definition see (2.5.31)) in the jj coupling scheme, i.e. the $(\frac{1}{2}l'j'||\mathbf{C}(k)||\frac{1}{2}lj)$, are obtained by reference to (5.4.6) and (7.1.8). Note that the formulas are independent of whether $l = j \pm \frac{1}{2}$.

(7.1.9)
$$(\tfrac{1}{2} l' \; l' \pm \tfrac{1}{2} ||\mathbf{C}(k)|| \tfrac{1}{2} \; l \; l \pm \tfrac{1}{2})$$
$$= 2 \cdot (-1)^{(j+k-j')/2} \left[\frac{(j'+j-k)!(j'+k-j)!(j+k-j')!}{(j'+j+k+1)!} \right]^{\frac{1}{2}}$$
$$\times \frac{\left(\dfrac{j+j'+k+1}{2}\right)!}{\left(\dfrac{j'+j-k-1}{2}\right)!\left(\dfrac{j+k-j'}{2}\right)!\left(\dfrac{j'+k-j}{2}\right)!}$$

(7.1.10)
$$(\tfrac{1}{2} l' \; l' \pm \tfrac{1}{2} ||\mathbf{C}(k)|| \tfrac{1}{2} \; l \; l \mp \tfrac{1}{2})$$
$$= 2 \cdot (-1)^{(j+k-j'-1)/2} \left[\frac{(j'+j-k)!(j'+k-j)!(j+k-j')!}{(j'+j+k+1)!} \right]^{\frac{1}{2}}$$
$$\times \frac{\left(\dfrac{j'+j+k}{2}\right)!}{\left(\dfrac{j'+j-k}{2}\right)!\left(\dfrac{j'+k-j-1}{2}\right)!\left(\dfrac{j+k-j'-1}{2}\right)!}$$

7.2. Selected Examples from Atomic, Molecular and Nuclear Physics

CENTRAL TWO-BODY INTERACTION. We consider the matrix elements of a central interaction between two particles in a scheme where the total angular momentum of the two particles is a good quantum number. The interaction is supposed to be some function $V(r_{12})$ of the distance r_{12} between the particles, whose position with respect to an origin O is given by two vectors $\mathbf{r}_1$ and $\mathbf{r}_2$.

---

[1] Cf. Racah (1942). If a fractional power of -1 should appear in evaluating these formulae it should be replaced by zero.

The expression

$$\frac{1}{r_{12}} = [r_1^2 + r_2^2 - 2r_1 r_2 \cos\theta_{12}]^{-\frac{1}{2}}$$

where θ_{12} is the angle between $\mathbf{r}_1$ and $\mathbf{r}_2$, may be developed in a series of Legendre polynomials (see (2.5.12))

$$\frac{1}{r_{12}} = \sum_{k=0}^{\infty} \frac{r_<^k}{r_>^{k+1}} P_k(\cos\theta_{12})$$

where $r_<$ is the lesser and $r_>$ the greater of r_1 and r_2. The interaction $V(r_{12})$ (of which $1/r_{12}$ is a special case—the electrostatic interaction[2]) may be developed in a similar series

$$V(r_{12}) = \sum_{k=0}^{\infty} (2k+1) V_k(r_1, r_2) P_k(\cos\theta_{12})$$

where

$$V_k(r_1, r_2) = \frac{1}{2}\int_0^{\pi} V(r_{12}) P_k(\cos\theta) \sin\theta \, d\theta$$

Expressions for the V_k in the case of typical nuclear interactions, e.g. the Gaussian

$$V(r_{12}) = -B \exp{-\left(\frac{r_{12}}{a}\right)^2}$$

and the Yukawa

$$V(r_{12}) = -\frac{B \exp{-r_{12}/a}}{r_{12}/a}$$

are given by Swiatecki (1951).

Now the quantity $P_k(\cos\theta_{12})$ is given in terms of the angles of the vectors $\mathbf{r}_1, \mathbf{r}_2$ by the spherical harmonic addition theorem (4.6.7) the right hand side of which may be considered as a scalar product of the tensors $\mathbf{C}^{(k)}(\theta_1 \varphi_1)$ and $\mathbf{C}^{(k)}(\theta_2 \varphi_2)$. The matrix element of $P_k(\cos\theta_{12})$ in the coupled scheme follows from (7.1.6) and (5.4.6)

$$(l_1' l_2' l' m' | P_k(\cos\theta_{12}) | l_1 l_2 l m)$$
$$= (l_1' l_2' l' m' | (\mathbf{C}_{(1)}^{(k)} \cdot \mathbf{C}_{(2)}^{(k)}) | l_1 l_2 l m)$$
$$= (-1)^{l_2 + l_2' + l} \delta_{ll'} \delta_{mm'} \begin{Bmatrix} l & l_2' & l_1' \\ k & l_1 & l_2 \end{Bmatrix} [(2l_1+1)(2l_1'+1)(2l_2+1)(2l_2'+1)]^{\frac{1}{2}}$$
$$\times \begin{pmatrix} l_1 & k & l_1' \\ 0 & 0 & 0 \end{pmatrix} \begin{pmatrix} l_2 & k & l_2' \\ 0 & 0 & 0 \end{pmatrix}$$

---

[2] Cf. Condon and Shortley (1935) p. 174.

which may be evaluated by reference to Table 5 (p. 130) and (3.8).

The radial part of the matrix element is expressed in terms of the *generalized Slater integral*

$$F^{(k)}(n_1\ l_1\ n_2\ l_2\ n_1'\ l_1'\ n_2'\ l_2')$$

$$= (2k+1) \int_0^\infty \int_0^\infty V_k(r_1, r_2) R_{n_1 l_1}(r_1) R_{n_2 l_2}(r_2) R_{n_1' l_1'}(r_1) R_{n_2' l_2'}(r_2) r_1^2 r_2^2 \, dr_1 \, dr_2$$

where $R_{nl}(r)$ is the radial part of the appropriate single particle eigenfunction. Methods for evaluating such integrals in nuclear problems are given by Swiatecki (1951) and Talmi (1952).

HYPERFINE STRUCTURE OF SYMMETRIC TOP MOLECULE.[3] The electrostatic interaction between a nucleus and the remainder of electrons and nuclei in the atom or molecule is given by

$$H_{el} = +\sum_{ip} \frac{e_i e_p}{|\mathbf{r}_i - \mathbf{r}_p|} = +\sum_{ipl} e_i e_p \frac{r_p^l}{r_i^{l+1}} P_l(\cos \theta_{ip})$$

where e_p is the charge of the pth proton with position vector $\mathbf{r}_p$ in the nucleus in question and e_i is the charge of the ith electron or proton with position vector $\mathbf{r}_i$ in the remainder of the atom or molecule. θ_{ip} is the angle between the vectors $\mathbf{r}_i$ and $\mathbf{r}_p$.

We consider the quadrupole term only in the multipole expansion on the right, and use the spherical harmonic addition theorem (4.6.7) to obtain

$$H_Q = \sum_{ipq} (-1)^q e_i e_p \frac{r_p^2}{r_i^3} C_q^{(2)}(\theta_i \varphi_i) C_{-q}^{(2)}(\theta_p \varphi_p) = (\mathbf{V} \cdot \mathbf{Q})$$

where

$$\mathbf{V} = \sum_i \frac{e_i}{r_i^3} \mathbf{C}^{(2)}(\theta_i \varphi_i), \qquad \mathbf{Q} = \sum_p e_p r_p^2 \mathbf{C}^{(2)}(\theta_p \varphi_p)$$

We define now the following relevant quantities:

I = spin angular momentum of nucleus in question.
J = angular momentum of the rest of the molecule (we assume the coupling between **I** and **J** weak in comparison with those couplings between the various angular momenta making up **J**).
K = component of **J** along the figure axis (z axis of moving frame).
M = component of **J** along the fixed z axis.
F = total molecular angular momentum (= **I** + **J**).

---

[3] The reader is referred to Ramsey (1953) for a detailed traditional treatment of this and allied problems. Octupole moment contributions to hyperfine structure are computed by Schwartz (1955).

The quadrupole splitting is then given by

$$\Delta E_Q = (\alpha\,J\,K,\beta\,I,F\,M_F|(\mathbf{V}\cdot\mathbf{Q})|\alpha\,J\,K,\beta\,I,F\,M_F)$$

$$= (-1)^{I+J+F}\begin{Bmatrix} F & I & J \\ 2 & J & I \end{Bmatrix}(\alpha\,J\,K||\mathbf{V}||\alpha\,J\,K)(\beta\,I||\mathbf{Q}||\beta\,I)$$

where α, β represent all relevant quantum numbers not related to angular momentum. The presence of the quantum number K should be noted; this is associated with the fact that the symmetric top eigenfunction is of the form[4]

$$R\mathfrak{D}_{MK}^{(J)}(\alpha\,\beta\,\gamma)\cdot(2J+1)^{\frac{1}{2}}$$

where $\alpha\,\beta\,\gamma$ are the Euler angles of the molecule and R is the scalar factor of the eigenfunction (cf. (4.8), Herzberg (1939)).

We refer to the previous example for the reduced matrix element of the nuclear quadrupole moment; we have

$$(\beta\,I||\mathbf{Q}||\beta\,I) = (-1)^{2I}\cdot\frac{e}{2}(\beta\,I\,I|\sum_p(3z_p^2 - r_p^2)|\beta\,I\,I)\bigg/\begin{pmatrix} I & 2 & I \\ -I & 0 & I \end{pmatrix}$$

$$= (2I+1)\left[\frac{2I+3}{I(2I-1)}\right]^{\frac{1}{2}}\cdot\frac{eQ}{2}$$

Use is made again of the spherical harmonic addition theorem to evaluate the z component of the tensor operator $\mathbf{V}$:

$$V_0 = \sum_i \frac{e_i}{r_i^3} C_0^{(2)}(\theta_i, \varphi_i)$$

$$= \sum_q (-1)^q C_q^{(2)}(\Theta, \Phi) \sum_i \frac{e_i}{r_i^3} C_{-q}^{(2)}(\Theta_i, \Phi_i)$$

where $\Theta, \Phi, \Theta_i, \Phi_i$ are the angles of the fixed z axis and the position vector of the ith electron respectively with respect to the frame of reference moving with the molecule.

Now we may evaluate the matrix element of $C_q^{(2)}(\Theta, \Phi)$ observing from the definition of the Euler angles (1.3) that $\Theta = \beta, \Phi = \pi - \gamma$.

We get accordingly

$$(\alpha\,J\,K'\,M'|C_q^{(2)}(\Theta, \Phi)|\alpha\,J\,K\,M)$$

$$= \frac{(2J+1)}{8\pi^2}\iiint \mathfrak{D}_{M'K'}^{(J)*}(\alpha\,\beta\,\gamma)C_q^{(2)}(\beta, \pi-\gamma)\mathfrak{D}_{MK}^{(J)}(\alpha\,\beta\,\gamma)\,d\alpha\,\sin\beta\,d\beta\,d\gamma$$

$$= \frac{(2J+1)}{8\pi^2}\iiint \mathfrak{D}_{M'K'}^{(J)*}(\alpha\,\beta\,\gamma)\mathfrak{D}_{0-q}^{(2)}(\alpha\,\beta\,\gamma)\mathfrak{D}_{MK}^{(J)}(\alpha\,\beta\,\gamma)\,d\alpha\,\sin\beta\,d\beta\,d\gamma$$

---

[4]Cf. (4.1) for normalization of the angular part.

where the relation (4.1.25) has been employed. The value of the integral is given by (4.2.7) and (4.6.2)

$$(\alpha\ J\ K'\ M'|C_q^{(2)}(\Theta, \Phi)|\alpha\ J\ K\ M)$$

$$= (-1)^{K'-M'}(2J+1)\begin{pmatrix} J & 2 & J \\ -M' & 0 & M \end{pmatrix}\begin{pmatrix} J & 2 & J \\ -K' & -q & K \end{pmatrix}$$

Since we consider only matrix elements diagonal in K we have necessarily $q = 0$.

It is convenient to express $\sum_i (e_i/r_i^3) C_0^{(2)}(\Theta_i, \Phi_i)$ in terms of the electrostatic potential V due to the remainder of electrons and protons of the molecule surrounding the nucleus with the quadrupole moment:

$$\sum_i \frac{e_i}{r_i^3} C_0^{(2)}(\Theta_i, \Phi_i) = \frac{1}{2}\frac{\partial^2 V}{\partial z'^2}$$

where the coordinate z' is in the frame of reference moving with the molecule.[5]

We obtain finally the reduced matrix element of the electron operator

$$(\alpha\ J\ K||\mathbf{V}||\alpha\ J\ K)$$

$$= \frac{1}{2}\left\langle\frac{\partial^2 V}{\partial z'^2}\right\rangle_\alpha (2J+1)(-1)^{J+K}\begin{pmatrix} J & 2 & J \\ -K & 0 & K \end{pmatrix}$$

$$= \frac{1}{2}\left\langle\frac{\partial^2 V}{\partial z'^2}\right\rangle_\alpha [3K^2 - J(J+1)]\left[\frac{2J+1}{(2J-1)(2J+3)(J)(J+1)}\right]^{\frac{1}{2}}$$

Evaluation of the 6-j symbol gives us the quadrupole splitting in terms of quantities supposed known:

$$\Delta E_Q = \frac{eQ}{2}\left\langle\frac{\partial^2 V}{\partial z'^2}\right\rangle_\alpha \frac{[3K^2 - J(J+1)][\frac{3}{4}C(C+1) - I(I+1)J(J+1)]}{I(2I-1)(2J-1)J(J+1)(2J+3)}$$

where $\quad C = F(F+1) - I(I+1) - J(J+1)$

This result corresponds to that of Ramsey (1953) p. 422. It may be shown by a lengthy calculation (Ramsey (1953) p. 373) that the quadrupole interaction may be expressed directly[6] in terms of the angular momentum operators **I** and **J**. The relevant factor is

$$3(\mathbf{I}\cdot\mathbf{J})^2 + \tfrac{3}{2}\mathbf{I}\cdot\mathbf{J} - \mathbf{I}^2\mathbf{J}^2$$

This expression may be evaluated to give the same result as obtained above.

---

[5]Cf. Ramsey (1953) p. 377.
[6]See also Schwinger (1952) Eq. 5.88.

118   7 · EVALUATION OF MATRIX ELEMENTS

MAGNETIC HYPERFINE STRUCTURE.[7] The interaction of a nuclear magnetic moment with the spin magnetic moment of an electron is a particular example of a *tensor* interaction, and is given by (cf. Kopfermann (1940))

$$H_M = -a_l \mathbf{I} \cdot \left\{ \mathbf{s} - \frac{3\mathbf{r}(\mathbf{r} \cdot \mathbf{s})}{r^2} \right\} \qquad (l \neq 0)$$

where $\mathbf{I}$ is the nuclear spin and $\mathbf{r}$ and $\mathbf{s}$ the position vector with respect to the nucleus and the spin vector respectively of the electron. The constant a_l is given by

$$a_l = R\alpha^2 a_0^3 \left(\frac{m}{M}\right) \left\langle \frac{1}{r^3} \right\rangle g(I)$$

where R = Rydberg constant
α = fine structure constant
a_0 = Bohr radius
m/M = electron-nucleon mass ratio
$g(I)$ = nuclear g factor.

We may form a tensor operator $\mathbf{X}$ of rank 1, namely

$$\mathbf{X}(1 q) = \sum_{q_1 q_2} s(1\ q_1) C_{q_2}^{(2)}(\theta, \varphi)(1\ q_1\ 2\ q_2 | 1\ 2\ 1\ q)$$

Then

$$\left\{ \mathbf{s} - \frac{3\mathbf{r}(\mathbf{r} \cdot \mathbf{s})}{r^2} \right\} = \sqrt{10}\ \mathbf{Y}$$

The factor $\sqrt{10}$ is easily obtained by computing the $q = 0$ component of $\mathbf{X}$ and comparing it with the z component of the left-hand side. Now let $\mathbf{J}$ be the total angular momentum of all the electrons in the atom and $\mathbf{F} = \mathbf{I} + \mathbf{J}$ be the total angular momentum of the atom. Then the net effect of the magnetic interaction between nucleus and electrons is

$$\Delta E_M = (\alpha\ J, I, F\ M_F | H_M | \alpha\ J, I, F\ M_F)$$

$$= (-1)^{I+J+F+1} a_l \cdot \sqrt{10}\ [(2I+1)I(I+1)]^{\frac{1}{2}} \begin{Bmatrix} F & I & J \\ 1 & J & I \end{Bmatrix} (\alpha\ J || \mathbf{X} || \alpha\ J)$$

$$= a_l \sqrt{10}\ \frac{I(I+1) + J(J+1) - F(F+1)}{[2J(2J+1)(2J+2)]^{\frac{1}{2}}} (\alpha\ J || \mathbf{X} || \alpha\ J)$$

where we have employed (7.1.7) and Table 5 and the reduced matrix element of $\mathbf{X}$ is supposed taken over all electrons. In *L-S* coupling the

---
[7] Cf. Trees (1953), Ramsey (1953).

diagonal reduced matrix element of X is given by (7.1.5):

$$(\alpha\,S\,L\,J||\mathsf{X}||\alpha\,S\,L\,J)$$
$$= \sqrt{3}\,(2J+1)\begin{Bmatrix} S & 1 & S \\ L & 2 & L \\ J & 1 & J \end{Bmatrix}(\alpha\,S||\mathsf{s}||\alpha\,S)(\alpha\,L||\mathsf{C}^{(2)}||\alpha\,L)$$

The reduced matrix elements on the right are given for 1-electron spectra by (5.4.5). Their evaluation for various configurations is discussed by Racah (1942, 1943). The 9-j symbol is computed by means of (6.4.17). We obtain, putting $\lambda = L$,

$$\begin{Bmatrix} S & 1 & S \\ L & 2 & L \\ J & 1 & J \end{Bmatrix} = \frac{\begin{Bmatrix} S & L & J \\ L & S & 1 \end{Bmatrix}\begin{Bmatrix} J & L & S \\ L & J & 1 \end{Bmatrix} + \dfrac{(-1)^{S+L+J+1}}{3(2L+1)}\begin{Bmatrix} S & J & L \\ J & S & 1 \end{Bmatrix}}{5\begin{Bmatrix} 2 & L & L \\ L & 1 & 1 \end{Bmatrix}}$$

INTENSITIES OF HYPERFINE TRANSITIONS FOR A SYMMETRIC TOP MOLECULE. In the rotating frame of coordinates only one component of the electric dipole moment $\mathbf{\mu}'$ has a nonzero expectation value, namely $\mu_0' = \mu_{z'}'$.

The first step is transformation to the fixed frame of reference (cf. (5.2.1)):

$$\langle \mu_\kappa \rangle = \langle \sum_\nu \mu_\nu' \mathfrak{D}_{\nu\kappa}^{(1)}(\alpha\,\beta\,\gamma)\rangle$$
$$= \langle \mu_0'\rangle \mathfrak{D}_{0\kappa}^{(1)}(\alpha\,\beta\,\gamma) = \langle \mu_0'\rangle C_\kappa^{(1)}(\beta,\gamma)$$

where we have used (4.1.25) and (2.5.31).

We sum over final states and polarizations to obtain

$$\text{Intensity} \sim \sum_{\kappa M_f} |(I\,J_i\,K_i\,F_i\,M_i|\mu_0' C_\kappa^{(1)}(\beta,\gamma)|I\,J_f\,K_f\,F_f\,M_f)|^2$$
$$= \frac{\mu_0'^2}{2F_i+1}|(I\,J_i\,K_i\,F_i||\mathsf{C}^{(1)}||I\,J_f\,K_f\,F_f)|^2$$

where the summation over the squares of 3-j symbols is given by (3.7.8). Application of (7.1.8) results in

$$\mu_0'^2(2F_f+1)\left|\begin{Bmatrix} J_i & F_i & I \\ F_f & J_f & 1 \end{Bmatrix}\right|^2 |(J_i\,K_i||\mathsf{C}^{(1)}||J_f\,K_f)|^2$$
$$= \mu_0'^2(2F_f+1)(2J_i+1)(2J_f+1)\left|\begin{Bmatrix} J_i & F_i & I \\ F_f & J_f & 1 \end{Bmatrix}\begin{pmatrix} J_i & 1 & J_f \\ -K_i & 0 & K_f \end{pmatrix}\right|^2$$

where the reduced matrix element of $\mathsf{C}^{(1)}$ has been evaluated by the method used in the previous example.

SUM RULE FOR TRANSITIONS IN L-S COUPLING. The total intensity of, say, a dipole transition in L-S coupling is given, following Condon and Shortley (1935) p. 238, by

$$S(\alpha\ S\ L\ J, \alpha'\ S\ L'\ J') = |(\alpha\ S\ L\ J||\mathbf{P}||\alpha'\ S\ L'\ J')|^2$$

which is shown by (7.1.8) to be equal to

$$(2J+1)(2J'+1)\left|\begin{Bmatrix} L & J & S \\ J' & L' & 1 \end{Bmatrix}\right|^2 |(\alpha\ L||\mathbf{P}||\alpha'\ L')|^2$$

The orthogonality of the 6-j symbols (6.2.9) furnishes the sum rule (cf. Condon and Shortley loc. cit.)

$$\sum_{J'} S(\alpha\ S\ L\ J, \alpha'\ S\ L'\ J') = \left(\frac{2J+1}{2L+1}\right)|(\alpha\ L||\mathbf{P}||\alpha'\ L')|^2$$

APPENDIX 1

Theorems Used in Chapter 3

The binomial coefficient

$$\binom{n}{r} \equiv \frac{n(n-1)\cdots(n-r+1)}{r!}$$

is given for positive integer n by $n!/r!(n-r)!$ and for negative integer $n = -\nu$, $(\nu > 0)$ by

$$(-1)^r \binom{\nu + r - 1}{r} = \frac{(-1)^r (\nu + r - 1)!}{r!(\nu - 1)!}$$

The addition theorem for the binomial coefficients,

$$\sum_\rho \binom{n}{\rho}\binom{m}{r-\rho} = \binom{n+m}{r}$$

is obtained by considering the coefficients of $x^r y^{n+m-r}$ on either side of the identity $(x+y)^n (x+y)^m \equiv (x+y)^{n+m}$. If we suppose n and m positive, we get immediately the relation

(A1.1) $\quad \sum_\rho [\rho!(m - r + \rho)!(n - \rho)!(r - \rho)!]^{-1} = \dfrac{(n+m)!}{n!m!r!(n+m-r)!}$

We set $\nu = -n > m \geq 0$ and obtain

$$\sum_\rho (-1)^\rho \frac{(\nu + \rho - 1)!}{\rho!(r - \rho)!(m - r + \rho)!} = \frac{(-1)^r (\nu - m + r - 1)!(\nu - 1)!}{r!(\nu - m - 1)!m!}.$$

On putting $p = \nu - 1$, $q = m - r$ we have

(A1.2) $\quad \sum_\rho (-1)^\rho \dfrac{(p+\rho)!}{\rho!(q+\rho)!(r-\rho)!} = \dfrac{(-1)^r p!(p-q)!}{r!(q+r)!(p-q-r)!}$,

if $q \geq 0$, $p - q \geq r \geq 0$.

In a similar way by putting $n = -\nu$, $m = -\mu$ we get after the replacements $\rho = c + \sigma$, $\nu + c - 1 = a$, $\mu + r - c - 1 = b$, $r - c = d$,

(A1.3) $\quad \sum_\sigma \dfrac{(a+\sigma)!(b-\sigma)!}{(c+\sigma)!(d-\sigma)!} = \dfrac{(a+b+1)!(a-c)!(b-d)!}{(c+d)!(a+b-c-d+1)!}$

APPENDIX 2

Approximate Expressions for Vector-Coupling Coefficients and 6-j Symbols[1]

The vector-coupling coefficient arising when one of the angular momenta involved is supposed small compared with the other two may be expressed approximately as a matrix element of a certain finite rotation. We must suppose also that the z components of the large angular momenta are large compared with the small one. The relation, using an obvious notation, is

(A2.1) $\quad (j\ m\ J\ M | j\ J\ J'\ m+M) \cong (-1)^{j-J+J'} d^{(j)}_{J'-J,\,m}(\theta)$

where

$$\cos\theta = \frac{M}{[J(J+1)]^{\frac{1}{2}}} \quad \text{and} \quad \sin\theta = \left[\frac{2(J+1)^2 - 2M^2}{(2J+1)(J+1)}\right]^{\frac{1}{2}}$$

When J is sufficiently large, we may of course write $\cos\theta = M/J$, etc. In giving the values to be inserted into the d function (cf. (4.1.11)) for $\cos\theta$ and $\sin\theta$, it has been assumed that j is an integer. Similar replacements may be made for $\cos\theta/2$ and $\sin\theta/2$ when this is not the case.

We take as a simple example

$$(1\ 1\ J\ M | 1J\ J+1\ M+1) = \left[\frac{J(J+1) + 2JM + M(M+1)}{(2J+1)(2J+2)}\right]^{\frac{1}{2}}$$

$$d^{(1)}_{11}(\theta) = \tfrac{1}{2}(1 + \cos\theta)$$

and see that J does not have to be very large for the agreement to be close.

The approximation problem for the 6-j symbols is more complicated, and not only since there is a greater choice of which arguments become large and which small. We shall consider one case which seems to be of interest in practice, namely when all the arguments except one are large. The corresponding approximate relation is

(A2.2) $\quad \begin{Bmatrix} J & J_2 & J_1 \\ j & J_1+\delta_1 & J_2+\delta_2 \end{Bmatrix} \cong \dfrac{(-1)^{J_1+J_2+J}}{[(2J_1+1)(2J_2+1)]^{\frac{1}{2}}} d^{(j)}_{-\delta_1,\,\delta_2}(\theta)$

---

[1] Cf. Brussaard and Tolhoek (1957) and Wigner (1959).

where

$$\cos\theta = \frac{J(J+1) - J_1(J_1+1) - J_2(J_2+1)}{2[J_1(J_1+1)J_2(J_2+1)]^{\frac{1}{2}}} \quad \text{(cf. (7.1))}$$

which corresponds to the 6-j symbol

$$\begin{Bmatrix} J & J_2 & J_1 \\ 1 & J_1 & J_2 \end{Bmatrix}.$$

We have a more complicated expression for $\sin\theta$ (see Table 5 for

$$\begin{Bmatrix} J & J_2 & J_1 \\ 1 & J_1-1 & J_2 \end{Bmatrix})$$

When $\delta_1 = \delta_2 = 0$ (and j = integer) we have

(A2.3) $$\begin{Bmatrix} J & J_2 & J_1 \\ j & J_1 & J_2 \end{Bmatrix} \cong \frac{(-1)^{J_1+J_2+J}}{[(2J_1+1)(2J_2+1)]^{\frac{1}{2}}} P_j(\cos\theta)$$

a result given by Racah (1951).

It is instructive to apply these relations to the formulas for the matrix elements of tensor operators given in Chapters 5 and 7; the correspondence between quantum mechanical and classical results is clearly demonstrated. Such correspondences are of importance in such problems as that of Coulomb excitation of nuclei, where, due to the long range of the Coulomb interaction, large orbital angular momenta are important.[2]

---

[2]Cf. Alder et al. (1956)

Table 1. Harmonic Polynomials and Spherical Harmonics.

$$\mathcal{Y}_{lm}(\mathbf{r}) = r^l Y_{lm}(\theta, \varphi)$$

l, m	$\mathcal{Y}_{lm}(\mathbf{r})$	$Y_{lm}(\theta, \varphi)$
0 0	$\dfrac{1}{2\sqrt{\pi}}$	$\dfrac{1}{2\sqrt{\pi}}$
1 0	$\dfrac{1}{2}\sqrt{\dfrac{3}{\pi}}\, z$	$\dfrac{1}{2}\sqrt{\dfrac{3}{\pi}}\, \cos\theta$
1 ±1	$\mp\dfrac{1}{2}\sqrt{\dfrac{3}{2\pi}}\,(x \pm iy)$	$\mp\dfrac{1}{2}\sqrt{\dfrac{3}{2\pi}}\,\sin\theta\, e^{\pm i\varphi}$
2 0	$\dfrac{1}{4}\sqrt{\dfrac{5}{\pi}}\,(2z^2 - x^2 - y^2)$	$\dfrac{1}{4}\sqrt{\dfrac{5}{\pi}}\,(2\cos^2\theta - \sin^2\theta)$
2 ±1	$\mp\dfrac{1}{2}\sqrt{\dfrac{15}{2\pi}}\, z(x \pm iy)$	$\mp\dfrac{1}{2}\sqrt{\dfrac{15}{2\pi}}\,\cos\theta\sin\theta\, e^{\pm i\varphi}$
2 ±2	$\dfrac{1}{4}\sqrt{\dfrac{15}{2\pi}}\,(x \pm iy)^2$	$\dfrac{1}{4}\sqrt{\dfrac{15}{2\pi}}\,\sin^2\theta\, e^{\pm 2i\varphi}$
3 0	$\dfrac{1}{4}\sqrt{\dfrac{7}{\pi}}\,(2z^2 - 3x^2 - 3y^2)z$	$\dfrac{1}{4}\sqrt{\dfrac{7}{\pi}}\,(2\cos^3\theta - 3\cos\theta\sin^2\theta)$
3 ±1	$\mp\dfrac{1}{8}\sqrt{\dfrac{21}{\pi}}\,(4z^2 - x^2 - y^2)(x \pm iy)$	$\mp\dfrac{1}{8}\sqrt{\dfrac{21}{\pi}}\,(4\cos^2\theta\sin\theta - \sin^3\theta)e^{\pm i\varphi}$
3 ±2	$\dfrac{1}{4}\sqrt{\dfrac{105}{2\pi}}\, z(x \pm iy)^2$	$\dfrac{1}{4}\sqrt{\dfrac{105}{2\pi}}\,\cos\theta\sin^2\theta\, e^{\pm 2i\varphi}$
3 ±3	$\mp\dfrac{1}{8}\sqrt{\dfrac{35}{\pi}}\,(x \pm iy)^3$	$\mp\dfrac{1}{8}\sqrt{\dfrac{35}{\pi}}\,\sin^3\theta\, e^{\pm 3i\varphi}$

Irreducible tensors containing in addition the components of some other vector $\mathbf{r}'$ may be constructed by polarization of the harmonics with the operator

$$\mathbf{r}' \cdot \nabla \equiv x'\frac{\partial}{\partial x} + y'\frac{\partial}{\partial y} + z'\frac{\partial}{\partial z}$$

Cf. Rose (1954).

Table 2.

$$\begin{pmatrix} j_1 & j_2 & j_3 \\ 0 & 0 & 0 \end{pmatrix} = (-1)^{\frac{1}{2}J} \left[\frac{(j_1+j_2-j_3)!(j_1+j_3-j_2)!(j_2+j_3-j_1)!}{(j_1+j_2+j_3+1)!} \right]^{\frac{1}{2}} \frac{(\frac{1}{2}J)!}{(\frac{1}{2}J-j_1)!(\frac{1}{2}J-j_2)!(\frac{1}{2}J-j_3)!}$$

if J is even.

$$\begin{pmatrix} j_1 & j_2 & j_3 \\ 0 & 0 & 0 \end{pmatrix} = 0 \quad \text{if} \quad J \text{ is odd} \quad \text{where} \quad J = j_1 + j_2 + j_3$$

$$\begin{pmatrix} J+\frac{1}{2} & J & \frac{1}{2} \\ M & -M-\frac{1}{2} & \frac{1}{2} \end{pmatrix} \qquad (-1)^{J-M-\frac{1}{2}} \left[\frac{J-M+\frac{1}{2}}{(2J+2)(2J+1)} \right]^{\frac{1}{2}} \qquad (J+\tfrac{1}{2},\, J,\, \tfrac{1}{2})$$

$$\begin{pmatrix} J+1 & J & 1 \\ M & -M-1 & 1 \end{pmatrix} \qquad (-1)^{J-M-1} \left[\frac{(J-M)(J-M+1)}{(2J+3)(2J+2)(2J+1)} \right]^{\frac{1}{2}}$$

$$\begin{pmatrix} J+1 & J & 1 \\ M & -M & 0 \end{pmatrix} \qquad (-1)^{J-M-1} \left[\frac{(J+M+1)(J-M+1)\cdot 2}{(2J+3)(2J+2)(2J+1)} \right]^{\frac{1}{2}} \qquad (J+1,\, J,\, 1)$$

$$\begin{pmatrix} J & J & 1 \\ M & -M-1 & 1 \end{pmatrix} \qquad (-1)^{J-M} \left[\frac{(J-M)(J+M+1)\cdot 2}{(2J+2)(2J+1)(2J)} \right]^{\frac{1}{2}}$$

$$\begin{pmatrix} J & J & 1 \\ M & -M & 0 \end{pmatrix} \qquad (-1)^{J-M} \frac{M}{[(2J+1)(J+1)J]^{\frac{1}{2}}} \qquad (J,\, J,\, 1)$$

Table 2 (continued)

$\begin{pmatrix} J+\frac{3}{2} & J & \frac{3}{2} \\ M & -M-\frac{3}{2} & \frac{3}{2} \end{pmatrix}$	$(-1)^{J-M+\frac{1}{2}}\left[\dfrac{(J-M-\frac{1}{2})(J-M+\frac{1}{2})(J-M+\frac{3}{2})}{(2J+4)(2J+3)(2J+2)(2J+1)}\right]^{\frac{1}{2}}$	
$\begin{pmatrix} J+\frac{3}{2} & J & \frac{3}{2} \\ M & -M-\frac{1}{2} & \frac{1}{2} \end{pmatrix}$	$(-1)^{J-M+\frac{1}{2}}\left[\dfrac{3(J-M+\frac{1}{2})(J-M+\frac{3}{2})(J+M+\frac{3}{2})}{(2J+4)(2J+3)(2J+2)(2J+1)}\right]^{\frac{1}{2}}$	$(J+\tfrac{3}{2}, J, \tfrac{3}{2})$
$\begin{pmatrix} J+\frac{1}{2} & J & \frac{3}{2} \\ M & -M-\frac{3}{2} & \frac{3}{2} \end{pmatrix}$	$(-1)^{J-M-\frac{1}{2}}\left[\dfrac{3(J-M-\frac{1}{2})(J-M+\frac{1}{2})(J+M+\frac{3}{2})}{(2J+3)(2J+2)(2J+1)2J}\right]^{\frac{1}{2}}$	
$\begin{pmatrix} J+\frac{1}{2} & J & \frac{3}{2} \\ M & -M-\frac{1}{2} & \frac{1}{2} \end{pmatrix}$	$(-1)^{J-M-\frac{1}{2}}\dfrac{J-M+\frac{1}{2}}{[(2J+3)(2J+2)(2J+1)2J]^{\frac{1}{2}}}(J+3M+\tfrac{3}{2})$	$(J+\tfrac{1}{2}, J, \tfrac{3}{2})$
$\begin{pmatrix} J+2 & J & 2 \\ M & -M-2 & 2 \end{pmatrix}$	$(-1)^{J-M}\left[\dfrac{(J-M-1)(J-M)(J-M+1)(J-M+2)}{(2J+5)(2J+4)(2J+3)(2J+2)(2J+1)}\right]^{\frac{1}{2}}$	
$\begin{pmatrix} J+2 & J & 2 \\ M & -M-1 & 1 \end{pmatrix}$	$2(-1)^{J-M}\left[\dfrac{(J+M+2)(J-M+2)(J-M+1)(J-M)}{(2J+5)(2J+4)(2J+3)(2J+2)(2J+1)}\right]^{\frac{1}{2}}$	$(J+2, J, 2)$
$\begin{pmatrix} J+2 & J & 2 \\ M & -M & 0 \end{pmatrix}$	$(-1)^{J-M}\left[\dfrac{6(J+M+2)(J+M+1)(J-M+2)(J-M+1)}{(2J+5)(2J+4)(2J+3)(2J+2)(2J+1)}\right]^{\frac{1}{2}}$	

TABLES

$(J+1, J, 2)$

$\begin{pmatrix} J+1 & J & 2 \\ M & -M-2 & 2 \end{pmatrix}$ $\quad 2(-1)^{J-M+1}\left[\dfrac{(J-M-1)(J-M)(J-M+1)(J+M+2)}{(2J+4)(2J+3)(2J+2)(2J+1)2J}\right]^{\frac{1}{2}}$

$\begin{pmatrix} J+1 & J & 2 \\ M & -M-1 & 1 \end{pmatrix}$ $\quad (-1)^{J-M+1}2(J+2M+2)\left[\dfrac{(J-M+1)(J-M)}{(2J+4)(2J+3)(2J+2)(2J+1)2J}\right]^{\frac{1}{2}}$

$\begin{pmatrix} J+1 & J & 2 \\ M & -M & 0 \end{pmatrix}$ $\quad (-1)^{J-M+1}2M\left[\dfrac{6(J+M+1)(J-M+1)}{(2J+4)(2J+3)(2J+2)(2J+1)2J}\right]^{\frac{1}{2}}$

$(J, J, 2)$

$\begin{pmatrix} J & J & 2 \\ M & -M-2 & 2 \end{pmatrix}$ $\quad (-1)^{J-M}\left[\dfrac{6(J-M-1)(J-M)(J+M+1)(J+M+2)}{(2J+3)(2J+2)(2J+1)(2J)(2J-1)}\right]^{\frac{1}{2}}$

$\begin{pmatrix} J & J & 2 \\ M & -M-1 & 1 \end{pmatrix}$ $\quad (-1)^{J-M}(1+2M)\left[\dfrac{6(J+M+1)(J-M)}{(2J+3)(2J+2)(2J+1)(2J)(2J-1)}\right]^{\frac{1}{2}}$

$\begin{pmatrix} J & J & 2 \\ M & -M & 0 \end{pmatrix}$ $\quad (-1)^{J-M}\dfrac{2[3M^2-J(J+1)]}{[(2J+3)(2J+2)(2J+1)(2J)(2J-1)]^{\frac{1}{2}}}$

Table 3. Prime factors of factorials.

n	2	3	5	7	11	13	17	19	23
5	3	1	1						
6	4	2	1						
7	4	2	1	1					
8	7	2	1	1					
9	7	4	1	1					
10	8	4	2	1					
11	8	4	2	1	1				
12	10	5	2	1	1				
13	10	5	2	1	1	1			
14	11	5	2	2	1	1			
15	11	6	3	2	1	1			
16	15	6	3	2	1	1			
17	15	6	3	2	1	1	1		
18	16	8	3	2	1	1	1		
19	16	8	3	2	1	1	1	1	
20	18	8	4	2	1	1	1	1	
21	18	9	4	3	1	1	1	1	
22	19	9	4	3	2	1	1	1	
23	19	9	4	3	2	1	1	1	1
24	22	10	4	3	2	1	1	1	1
25	22	10	6	3	2	1	1	1	1

Example: $6! = 2^4 \cdot 3^2 \cdot 5$

Table 4.

$$\mathfrak{D}^{(j)}_{m'm}\left(0\,\frac{\pi}{2}\,0\right) \equiv d^{(j)}_{m'm}\left(\frac{\pi}{2}\right) \equiv \Delta^{(j)}_{m'm}$$

$(j = \tfrac{1}{2})$

$m'\backslash m$	$+\tfrac{1}{2}$	$-\tfrac{1}{2}$
$+\tfrac{1}{2}$	$\tfrac{1}{\sqrt{2}}$	$\tfrac{1}{\sqrt{2}}$
$-\tfrac{1}{2}$	$-\tfrac{1}{\sqrt{2}}$	$\tfrac{1}{\sqrt{2}}$

$(j = 1)$

$m'\backslash m$	$+1$	0	-1
$+1$	$\tfrac{1}{2}$	$\tfrac{1}{\sqrt{2}}$	$\tfrac{1}{2}$
0	$-\tfrac{1}{\sqrt{2}}$	0	$\tfrac{1}{\sqrt{2}}$
-1	$\tfrac{1}{2}$	$-\tfrac{1}{\sqrt{2}}$	$\tfrac{1}{2}$

$(j = \tfrac{3}{2})$

$m'\backslash m$	$+\tfrac{3}{2}$	$+\tfrac{1}{2}$	$-\tfrac{1}{2}$	$-\tfrac{3}{2}$
$+\tfrac{3}{2}$	$\tfrac{1}{2\sqrt{2}}$	$\tfrac{\sqrt{3}}{2\sqrt{2}}$	$\tfrac{\sqrt{3}}{2\sqrt{2}}$	$\tfrac{1}{2\sqrt{2}}$
$+\tfrac{1}{2}$	$-\tfrac{\sqrt{3}}{2\sqrt{2}}$	$-\tfrac{1}{2\sqrt{2}}$	$\tfrac{1}{2\sqrt{2}}$	$\tfrac{\sqrt{3}}{2\sqrt{2}}$
$-\tfrac{1}{2}$	$\tfrac{\sqrt{3}}{2\sqrt{2}}$	$-\tfrac{1}{2\sqrt{2}}$	$-\tfrac{1}{2\sqrt{2}}$	$\tfrac{\sqrt{3}}{2\sqrt{2}}$
$-\tfrac{3}{2}$	$-\tfrac{1}{2\sqrt{2}}$	$\tfrac{\sqrt{3}}{2\sqrt{2}}$	$-\tfrac{\sqrt{3}}{2\sqrt{2}}$	$\tfrac{1}{2\sqrt{2}}$

$(j = 2)$

$m'\backslash m$	$+2$	$+1$	0	-1	-2
$+2$	$\tfrac{1}{4}$	$\tfrac{1}{2}$	$\tfrac{\sqrt{3}}{2\sqrt{2}}$	$\tfrac{1}{2}$	$\tfrac{1}{4}$
$+1$	$-\tfrac{1}{2}$	$-\tfrac{1}{2}$	0	$\tfrac{1}{2}$	$\tfrac{1}{2}$
0	$\tfrac{\sqrt{3}}{2\sqrt{2}}$	0	$-\tfrac{1}{2}$	0	$\tfrac{\sqrt{3}}{2\sqrt{2}}$
-1	$-\tfrac{1}{2}$	$\tfrac{1}{2}$	0	$-\tfrac{1}{2}$	$\tfrac{1}{2}$
-2	$\tfrac{1}{4}$	$-\tfrac{1}{2}$	$\tfrac{\sqrt{3}}{2\sqrt{2}}$	$-\tfrac{1}{2}$	$\tfrac{1}{4}$

Table 5. Formulas for the 6-j symbol.

$$\begin{Bmatrix} a & b & c \\ 1 & c-1 & b-1 \end{Bmatrix} = (-1)^s \left[\frac{s(s+1)(s-2a-1)(s-2a)}{(2b-1)2b(2b+1)(2c-1)2c(2c+1)} \right]^{\frac{1}{2}}$$

$$\begin{Bmatrix} a & b & c \\ 1 & c-1 & b \end{Bmatrix} = (-1)^s \left[\frac{2(s+1)(s-2a)(s-2b)(s-2c+1)}{2b(2b+1)(2b+2)(2c-1)2c(2c+1)} \right]^{\frac{1}{2}}$$

$$\begin{Bmatrix} a & b & c \\ 1 & c-1 & b+1 \end{Bmatrix} = (-1)^s \left[\frac{(s-2b-1)(s-2b)(s-2c+1)(s-2c+2)}{(2b+1)(2b+2)(2b+3)(2c-1)2c(2c+1)} \right]^{\frac{1}{2}}$$

$$\begin{Bmatrix} a & b & c \\ 1 & c & b \end{Bmatrix} = (-1)^{s+1} \frac{2[b(b+1)+c(c+1)-a(a+1)]}{[2b(2b+1)(2b+2)2c(2c+1)(2c+2)]^{\frac{1}{2}}}$$

where $s = a + b + c$.

$$\begin{Bmatrix} a & b & c \\ \tfrac{3}{2} & c-\tfrac{3}{2} & b-\tfrac{3}{2} \end{Bmatrix} = (-1)^s \left[\frac{(s-1)s(s+1)(s-2a-1)(s-2a-2)(s-2a)}{(2b-2)(2b-1)2b(2b+1) \cdot (2c-2)(2c-1)2c(2c+1)} \right]^{\frac{1}{2}}$$

$$\begin{Bmatrix} a & b & c \\ \tfrac{3}{2} & c-\tfrac{3}{2} & b-\tfrac{1}{2} \end{Bmatrix} = (-1)^s \left[\frac{3s(s+1)(s-2a-1)(s-2a)(s-2b)(s-2c+1)}{(2b-1)2b(2b+1)(2b+2) \cdot (2c-2)(2c-1)2c(2c+1)} \right]^{\frac{1}{2}}$$

$$\begin{Bmatrix} a & b & c \\ \tfrac{3}{2} & c-\tfrac{3}{2} & b+\tfrac{1}{2} \end{Bmatrix} = (-1)^s \left[\frac{3(s+1)(s-2a)(s-2b-1)(s-2b)(s-2c+1)(s-2c+2)}{2b(2b+1)(2b+2)(2b+3) \cdot (2c-2)(2c-1)2c(2c+1)} \right]^{\frac{1}{2}}$$

TABLES   131

$$\begin{Bmatrix} a & b & c \\ \tfrac{3}{2} & c-\tfrac{3}{2} & b+\tfrac{3}{2} \end{Bmatrix} = (-1)^s \left[\frac{(s-2b-2)(s-2b-1)(s-2b)(s-2c+1)(s-2c+2)(s-2c+3)}{(2b+1)(2b+2)(2b+3)(2b+4)\cdot(2c-2)(2c-1)2c(2c+1)}\right]^{\tfrac{1}{2}}$$

$$\begin{Bmatrix} a & b & c \\ \tfrac{3}{2} & c-\tfrac{1}{2} & b-\tfrac{1}{2} \end{Bmatrix} = (-1)^s \frac{[2(s-2b)(s-2c)-(s+2)(s-2a-1)][(s+1)(s-2a)]^{\tfrac{1}{2}}}{[(2b-1)2b(2b+1)(2b+2)\cdot(2c-1)2c(2c+1)(2c+2)]^{\tfrac{1}{2}}}$$

$$\begin{Bmatrix} a & b & c \\ \tfrac{3}{2} & c-\tfrac{1}{2} & b+\tfrac{1}{2} \end{Bmatrix} = (-1)^s \frac{[(s-2b-1)(s-2c)-2(s+2)(s-2a)][(s-2b)(s-2c+1)]^{\tfrac{1}{2}}}{[2b(2b+1)(2b+2)(2b+3)\cdot(2c-1)2c(2c+1)(2c+2)]^{\tfrac{1}{2}}}$$

where $s = a+b+c$.

$$\begin{Bmatrix} a & b & c \\ 2 & c-2 & b-2 \end{Bmatrix} = (-1)^s \left[\frac{(s-2)(s-1)s(s+1)\cdot(s-2a-2)(s-2a-1)(s-2a)}{(2b-3)(2b-2)(2b-1)2b(2b+1)\cdot(2c-3)(2c-2)(2c-1)2c(2c+1)}\right]^{\tfrac{1}{2}}$$

$$\begin{Bmatrix} a & b & c \\ 2 & c-2 & b-1 \end{Bmatrix} = (-1)^s \cdot 2\cdot \left[\frac{(s-1)s(s+1)\cdot(s-2a-2)(s-2a-1)(s-2a)(s-2b)(s-2c+1)}{(2b-2)(2b-1)2b(2b+1)(2b+2)\cdot(2c-3)(2c-2)(2c-1)2c(2c+1)}\right]^{\tfrac{1}{2}}$$

$$\begin{Bmatrix} a & b & c \\ 2 & c-2 & b \end{Bmatrix} = (-1)^s \left[\frac{6s(s+1)\cdot(s-2a-1)(s-2a)(s-2b-1)(s-2b)(s-2c+1)(s-2c+2)}{(2b-1)2b(2b+1)(2b+2)(2b+3)\cdot(2c-3)(2c-2)(2c-1)2c(2c+1)}\right]^{\tfrac{1}{2}}$$

$$\begin{Bmatrix} a & b & c \\ 2 & c-2 & b+1 \end{Bmatrix} = (-1)^s \cdot 2\cdot \left[\frac{(s+1)(s-2a)(s-2b-2)(s-2b-1)(s-2b)(s-2c+1)(s-2c+2)(s-2c+3)}{2b(2b+1)(2b+2)(2b+3)(2b+4)\cdot(2c-3)(2c-2)(2c-1)2c(2c+1)}\right]^{\tfrac{1}{2}}$$

Table 5 (Continued)

$$\begin{Bmatrix} a & b & c \\ 2 & c-2 & b+2 \end{Bmatrix}$$
$$= (-1)^s \left[\frac{(s-2b-3)(s-2b-2)(s-2b-1)(s-2b)\cdot(s-2c+1)(s-2c+2)(s-2c+3)(s-2c+4)}{(2b+1)(2b+2)(2b+3)(2b+4)(2b+5)\cdot(2c-3)(2c-2)(2c-1)2c(2c+1)} \right]^{\frac{1}{2}}$$

$$\begin{Bmatrix} a & b & c \\ 2 & c-1 & b-1 \end{Bmatrix} = (-1)^s \cdot \frac{4[(a+b)(a-b+1)-(c-1)(c-b+1)][s(s+1)(s-2a-1)(s-2a)]^{\frac{1}{2}}}{[(2b-2)(2b-1)2b(2b+1)(2b+2)\cdot(2c-2)(2c-1)2c(2c+1)(2c+2)]^{\frac{1}{2}}}$$

$$\begin{Bmatrix} a & b & c \\ 2 & c-1 & b \end{Bmatrix} = (-1)^s \cdot 2 \frac{[(a+b+1)(a-b)-c^2+1][6(s+1)(s-2a)(s-2b)(s-2c+1)]^{\frac{1}{2}}}{[(2b-1)2b(2b+1)(2b+2)(2b+3)\cdot(2c-2)(2c-1)2c(2c+1)(2c+2)]^{\frac{1}{2}}}$$

$$\begin{Bmatrix} a & b & c \\ 2 & c-1 & b+1 \end{Bmatrix}$$
$$= (-1)^s \cdot \frac{4[(a+b+2)(a-b-1)-(c-1)(b+c+2)][(s-2b-1)(s-2b)(s-2c+1)(s-2c+2)]^{\frac{1}{2}}}{[2b(2b+1)(2b+2)(2b+3)(2b+4)\cdot(2c-2)(2c-1)2c(2c+1)(2c+2)]^{\frac{1}{2}}}$$

$$\begin{Bmatrix} a & b & c \\ 2 & c & b \end{Bmatrix} = (-1)^s \cdot \frac{2[3X(X-1) - 4b(b+1)c(c+1)]}{[(2b-1)2b(2b+1)(2b+2)(2b+3)\cdot(2c-1)2c(2c+1)(2c+2)(2c+3)]^{\frac{1}{2}}}$$

where $s = a + b + c$, $\quad X = b(b+1) + c(c+1) - a(a+1)$.

CITED REFERENCES

Alder, K. Beiträge zur Theorie der Richtungskorrelation. *Helv. Phys. Acta 25*, 235–258 (1952).

Alder K., Bohr A., Huus T., Mottelson B., and Winther A. Study of Nuclear Structure by Electromagnetic Excitation with Accelerated Ions *Revs. Mod. Phys. 28*, 432 (1956).

Bailey, W. N. *Generalised Hypergeometric Series*. Cambridge, 1935.

Bateman, H. *Partial Differential Equations of Mathematical Physics*. Cambridge, 1932.

Bauer, E. *Introduction à la théorie des groupes et ses applications à la physique quantique*. Presses Universitaires de France, 1933.

Bethe, H. A. *Handbuch der Physik 24*. Springer 1933.

Biedenharn, L. C. *Tables of the Racah Coefficients*. Oak Ridge National Laboratory, Physics Division, ORNL-1098, April 1952.

Biedenharn, L. C. *J. Math. and Phys. 31*, 287 (1953).

Biedenharn, L. C., and Simon, A. *Revised Z Tables of the Racah Coefficients*. Oak Ridge National Laboratory, Physics Division, ORNL-1501, suppl. 1, February 1954.

Biedenharn, L. C., Blatt, J. M., and Rose, M. E. Some properties of the Racah and associated coefficients. *Revs. Mod. Phys. 24*, 249–257 (1952).

Biedenharn, L. C., and Rose, M. E. Theory of angular correlations of nuclear radiations. *Revs. Mod. Phys. 25*, 729–777 (1953).

Blatt, J. M., and Weisskopf, V. F. *Theoretical Nuclear Physics*. Wiley, 1952.

Bloch, F. *Phys. Rev. 62*, 305 (A) (1942).

Bohr, A. *Dan. Mat. Fys. Medd. 26 (14)*, 1952.

Bohr, A., and Mottelson, B. R. *Dan. Mat. Fys. Medd. 27 (16)*, 1953.

Bohr, A., and Mottelson, B. R. *Dan. Mat. Fys. Medd. 30 (1)*, 1955.

Bohr, N. *Phil. Mag. 26*, 1 (1913).

Bopp, F., and Haag, R. *Z. Naturforsch. 5a*, 644 (1950).

Born, M., and Jordan, P. *Elementare Quantenmechanik*. Springer, 1930.

Boys, S. F. Electronic wave functions. IV. Some general theorems for the calculation of Schrödinger integrals between complicated vector-coupled functions for many-electron atoms. *Proc. Roy. Soc. A207*, 181–197 (1951).

Brinkman, H. C. *Applications of Spinor Invariants in Atomic Physics*. North-Holland, 1956.

Cartan, E. *Leçons sur la théorie des Spineurs*. Hermann et Cie., Paris, 1938.

Casimir, H. B. G. *Rotation of a Rigid Body in Quantum Mechanics*. Thesis, Leyden, 1931.

Coester, F., and Jauch, J. M. *Helv. Phys. Acta.* **26**, 3 (1953).
Condon, E. U., and Shortley, G. H. *The Theory of Atomic Spectra.* Cambridge, 1935.
Corben, H. C., and Schwinger, J. S. *Phys. Rev.* **58**, 953 (1940).
Darwin, C. G. *Proc. Roy. Soc.* **A118**, 654 (1928).
Dennison, D. M. *Revs. Mod. Phys.* **3**, 310 (1931).
Dirac, P. A. M. *The Principles of Quantum Mechanics.* Oxford, 1947.
Eckart, C. The application of group theory to the quantum dynamics of monatomic systems. *Revs. Mod. Phys.* **2**, 305–380 (1930).
Edmonds, A. R., and Flowers, B. H. Studies in jj-coupling. II. Fractional parentage coefficients and the central force energy matrix for equivalent particles. *Proc. Roy. Soc.* **A214**, 515–530 (1952).
Elliott, J. P. Theoretical studies in nuclear structure. V. The matrix elements of non-central forces with an application to the $2p$-shell. *Proc. Roy. Soc.* **A218**, 345–370 (1953).
Falkoff, D. L., and Uhlenbeck, G. E. *Phys. Rev.* **79**, 323 (1950).
Falkoff, D. L., Colladay, G. S., and Sells, R. E. Transformation amplitudes for vector addition of angular momentum; $(j3mm'|\, j3\, JM)$. *Can. J. Phys.* **30**, 253–256 (1952).
Fano, U. Statistical matrix techniques and their application to the directional correlations of radiations. *U. S. Natl. Bur. Standards Rept. 1214*. (Unpublished)
Franz, W. *Z. Physik* **127**, 363 (1950).
Goldstein, H. *Classical Mechanics.* Addison-Wesley, 1950.
De Groot, S. R., and Tolhoek, H. A. *Beta- and Gamma-Ray Spectroscopy* (ed. K. Siegbahn). North-Holland, 1955.
Güttinger, P. *Z. Physik* **73**, 169 (1932).
Güttinger, P., and Pauli, W. *Z. Physik* **67**, 743 (1931).
Hahn, W. *Math. Nachr.* **2**, 4 (1949).
Herzberg, G. *Molecular Spectra and Molecular Structure.* Prentice-Hall, 1939.
Hobson, E. W. *The Theory of Spherical and Ellipsoidal Harmonics.* Cambridge, 1931.
Hope, J. Ph.D. Thesis. London, 1951.
Hund, F. *Z. Physik* **51**, 1 (1928).
Jahn, H. A. Theoretical studies in nuclear structure. II. Nuclear d^2, d^3, and d^4 configurations. Fractional parentage coefficients and central force matrix elements. *Proc. Roy. Soc.* **A205**, 192–237 (1951).
Jahn, H. A., and Hope, J. Symmetry properties of the Wigner 9-j symbol. *Phys. Rev.* **93**, 318–321 (1954).
Jahnke, E., and Emde, F. *Funktionentafeln.* Teubner, 1933.
Jordan, C. *Calculus of Finite Differences.* Chelsea, 1947.
Kemble, E. C. *The Fundamental Principles of Quantum Mechanics.* McGraw-Hill, 1937.

King, G. W., Hainer, R. M., and Cross, P. C. *J. Chem. Phys. 11*, 27 (1943).

Klein, F., and Sommerfeld, A. *Theorie des Kreisels*. Teubner, 1914.

Kopfermann, H. *Kernmomente*. Akad. Verlagsgesellschaft, 1940.

Kramers, H. A. *Proc. Roy. Soc. Amsterdam 33*, 953 (1930); *34*, 956 (1931).

Kronig, R. de L., and Rabi, I. I. *Phys. Rev. 29*, 262 (1927).

Landau, L., and Lifschitz, E. *Kvantovaya Mekhanika*. Moscow, 1948.

Landé, A. *Z. Physik 15*, 189 (1923).

Littlewood, D. E. *Theory of Group Characters*. Oxford, 1950.

Milne-Thomson, L. M. *Calculus of Finite Differences*. Macmillan, 1933.

Morse, P. M., and Feshbach, H. *Methods of Theoretical Physics*. McGraw-Hill, 1953.

Mulliken, R. S. *Phys. Rev. 59*, 873 (1941).

Neumann, J. von. *Göttinger Nachr.*, 245 (1927).

Neumann, J. von. *Mathematische Grundlagen der Quantenmechanik*. Springer, 1932.

Ord-Smith, R. J. The symmetry relations of the 12-j symbol. *Phys. Rev. 94*, 1227–1228 (1954).

Pauli, W. *Z. Physik 43*, 601 (1927).

Pauli, W. *Revs. Mod. Phys. 13*, 203 (1941).

Pauli, W. *Lecture Notes*. École d'Éte, Les Houches, 1952.

Racah, G. Theory of complex spectra II. *Phys. Rev. 62*, 438 (1942).

Racah, G. Theory of complex spectra III. *Phys. Rev. 63*, 367 (1943).

Racah, G. *Lectures on Group Theory*. Inst. for Adv. Study, Princeton, 1951.

Ramsey, N. F. *Experimental Nuclear Physics*, Vol. I, Part III (ed. E. Segré). Wiley, 1953.

Reiche, F., and Rademacher, H. Die Quantelung des symmetrischen Kreisels nach Schrödingers Undulationsmechanik. *Z. Physik 39*, 444–464 (1926).

Rose, M. E. *Proc. Phys. Soc. 67A*, 239 (1954).

Rose, M. E. *Multipole Fields*. Wiley, 1955.

Rose, M. E., and Osborn, R. K. Nuclear matrix elements in β decay. *Phys. Rev. 93*, 1326–1336 (1954).

Sato, M. *Progr. Theor. Phys. 13*, 405 (1955).

Saito, R., and Morita, M. *Progr. Theor. Phys. 13*, 540 (1955).

Schiff, L. *Quantum Mechanics*. McGraw-Hill, 1949.

Schwartz, C. *Phys. Rev. 97*, 380 (1955).

Schwinger, J. *On Angular Momentum*. U. S. Atomic Energy Commission, NYO-3071, 1952.

de Shalit, A. The energy levels of odd-odd nuclei. *Phys. Rev. 91*, 1479–1486 (1953).

Sharp, W. T., et al. Tables of coefficients for angular distribution analysis. *CRT-556, AECL No. 97, Atomic Energy of Canada Ltd.*, 1954.
Shortley, G. H., and Fried, B. *Phys. Rev.* **54**, 739 (1938).
Simon, A. Theory of polarized particles and gamma rays in nuclear reactions. *Phys. Rev.* **92**, 1050–1060 (1953).
Simon, A. *Numerical Table of the Clebsch-Gordan Coefficients.* Oak Ridge National Laboratory, Physics Division, ORNL-1718, June 1954.
Simon, A., Vander Sluis, J. H., and Biedenharn, L. C. *Tables of the Racah Coefficients.* Oak Ridge National Laboratory, Physics Division, ORNL-1679, March 1954.
Simon, A., and Welton, T. A. Production of polarized particles in nuclear reactions. *Phys. Rev.* **90**, 1036–1043 (1953). Errata. *Phys. Rev.* **93**, 1435 (1954).
Sommerfeld, A. *Ann. Physik* **51**, 1 (1916).
Sommerfeld, A. *Atombau und Spektrallinien.* Vieweg, 1939.
Swiatecki, W. J. *Proc. Roy. Soc.* **A205**, 238 (1951).
Synge, J. L., and Griffith, B. A. *Principles of Mechanics.* McGraw-Hill, 1949.
Szegö, G. *Orthogonal Polynomials.* New York, Amer. Math. Soc., 1939 (Colloquium Publications Vol. 23).
Talmi, I. *Helv. Phys. Acta* **25**, 185 (1952).
Talmi, I. The evaluation of the energy matrix of the tensor forces. *Phys. Rev.* **89**, 1065–1071 (1953).
Tolman, R. C. *Principles of Statistical Mechanics.* Oxford, 1938.
Trees, R. E. *Phys. Rev.* **92**, 308 (1953).
Tricomi, F. G. *Vorlesungen über Orthogonalreihen.* Springer, 1955.
van der Waerden, B. L. *Die Gruppentheoretische Methode in der Quantenmechanik*, Springer, 1931.
van der Waerden, B. L. *Moderne Algebra.* Springer, 1950.
Watson, G. N. *Bessel Functions.* Cambridge, 1944.
Weyl, H. *Group Theory and Quantum Mechanics.* Methuen, 1931.
Weyl, H. *The Classical Groups.* Princeton, 1939.
Whittaker, E. *Analytical Dynamics.* Cambridge, 1917.
Whittaker, E., and Watson, G. N. *Modern Analysis.* Cambridge, 1946.
Wigner, E. P. *Z. Physik* **43**, 624 (1927).
Wigner, E. P. *Gruppentheorie.* Vieweg, 1931.
Wigner, E. P. *Göttinger Nachr.*, 546 (1932).
Wigner, E. P. On matrices which reduce Kronecker products of representations of S. R. groups. 1951 (unpublished).
van Winter, C. *Physica* **20**, 274 (1954).

ADDITIONAL REFERENCES

Wigner, E. P. *Group Theory*. Academic Press, 1959.

Fano, U. and Racah, G. *Irreducible Tensorial Sets.* Academic Press, 1959.

Smith, K. *Table of Wigner 9-j symbols for integral and half-integral values of the parameters*. ANL-5860, Argonne National Laboratory, Chicago, 1958.

Howell, K. M. *Tables of Wigner 6-j symbols*. Research Report 58-1, University of Southampton Mathematics Department, 1958.

Howell, K. M. *Revised Tables of 6-j symbols*. Research Report 59-1, University of Southampton Mathematics Department, 1959.

Jahn, H. A. and Howell, K. M. *New Parameters and Symmetry Relations for the Wigner 6-j symbol*. Research Report 59-1A, University of Southampton Mathematics Department, 1959.

Howell, K. M. *Tables of 9-j symbols*. Research Report 59-2, University of Southampton Mathematics Department, 1959.

Centre National d'Études des Télécommunications. *Tables des Fonctions de Legendre Associées*. Éditions de la Revue d'Optique, Paris, 1952.

Innes, F. R. and Ufford, C. W. *Phys. Rev.* 111, 194, 1958.

Rotenberg, M., Bivins, R., Metropolis, N., and Wooten, J. K. *The 3-j and 6-j Symbols*. Technology Press, M.I.T., 1959.

Erdélyi, A. (ed.). *Higher Transcendental Functions*. McGraw-Hill, 1953.

Brussaard, P. J. and Tolhoek, H. A. *Physica* 23, 955, 1957.

BIBLIOGRAPHY

Applications of the methods discussed in this book may be found in the following works. The bibliography is not claimed to be exhaustive; however, the papers marked by an asterisk (*) contain collections of references in the respective fields.

(i) Atomic spectra (see also (vi)).

> Bernal, M. J. M., and Boys, S. F. Electronic wave functions. VII: Methods of evaluating the fundamental coefficients for the expansion of vector-coupled Schrödinger integrals and some values of these. *Phil. Trans. Roy. Soc.* A245, 116–138 (1952).
>
> Boys, S. F. Electronic wave functions. V: Systematic reduction methods for all Schrödinger integrals of conventional systems of antisymmetric vector-coupled functions. *Proc. Roy. Soc.* A207, 197–215 (1951).
>
> Innes, F. R. Matrix elements of spin-spin interaction. *Phys. Rev.* 91, 31–34 (1953).
>
> Racah, G. Theory of complex spectra. II. *Phys. Rev.* 62, 438 (1942).
>
> Racah, G. Theory of complex spectra. III. *Phys. Rev.* 63, 367 (1943).
>
> Racah, G. Theory of complex spectra. IV. *Phys. Rev.* 76, 1352–1365 (1949).
>
> Rohrlich, F. On the configurations d^2sp, d^3p, and d^7p. *Phys. Rev.* 74, 1372–1380 (1948).
>
> Rohrlich, F. The classification of the odd terms of TiI. *Phys. Rev.* 74, 1381–1396 (1948).
>
> Rosenzweig, N. The configuration interaction between the odd terms in the iron group. *Phys. Rev.* 88, 580–586 (1952).
>
> Trees, R. E. Hyperfine structure formulas for LS coupling. *Phys. Rev.* 92, 308–314 (1953).

(ii) β decay

> Biedenharn, L. C., and Rose, M. E. Theory of angular correlations of nuclear radiations. *Revs. Mod. Phys.* 25, 729–777 (1953).*
>
> Rose, M. E., and Osborn, R. K. Nuclear matrix elements in β-decay. *Phys. Rev.* 93, 1326–1336 (1954).
>
> Talmi, I. Matrix elements of β-decay in jj-coupling. *Phys. Rev.* 91, 122–125 (1953).

(iii) γ *emission*

Moszkowski, S. A. Theory of multipole radiation. Siegbahn, K. (ed.). *Beta and Gamma Ray Spectroscopy*, pp. 373–395. North-Holland, 1955.

Rose, M. E. *Multipole Fields*. Wiley, 1955*

(iv) *Directional correlation of nuclear radiations*

Biedenharn, L. C., and Rose, M. E. Theory of angular correlations of nuclear radiations. *Revs. Mod. Phys.* 25, 729–777 (1953).*

Fano, U. Statistical matrix techniques and their applications to the directional correlations of radiations. *U. S. Natl. Bur. Standards Rept. 1214*. (Unpublished)

Frauenfelder, H. Angular correlation. Siegbahn, K. (ed.). *Beta and Gamma Ray Spectroscopy*, pp. 531–599. North-Holland, 1955.*

de Groot, S. R., and Tolhoek, H. A. Theory of angular effects of radiations from oriented nuclei. Siegbahn, K. (ed.). *Beta and Gamma Ray Spectroscopy*, pp. 613–623. North-Holland, 1955.

Sharp, W. T., et al. *Tables of Coefficients for Angular Distribution Analysis*. CRT-556, AECL No. 97, Atomic Energy Commission of Canada Ltd., 1954.

(v) *Collective model of the nucleus*

Bohr, A. The coupling of nuclear surface oscillations to the motion of individual nucleons. *Dan. Mat. Fys. Medd.* 26 (14), 1–40 (1952).

Bohr, A., and Mottelson, B. R. Collective and individual particle aspects of nuclear structure. *Dan. Mat. Fys. Medd.* 27 (16), 1–174 (1953).

Bohr, A., and Mottelson, B. R. Collective nuclear motion and the unified model. Siegbahn, K. (ed.). *Beta and Gamma Ray Spectroscopy*, pp. 468–493. North-Holland, 1955.*

(vi) *Fractional parentage coefficient methods*

The papers marked by a dagger (†) deal mainly with the theory of fractional parentage coefficients and of their calculation.

Edmonds, A. R., and Flowers, B. H. Studies in jj-coupling II. Fractional parentage coefficients and the central force energy matrix for equivalent particles. *Proc. Roy. Soc.* A214, 515–530 (1952).†

Edmonds, A. R., and Flowers, B. H. Studies in jj-coupling. III. Nuclear energy levels. *Proc. Roy. Soc.* A215, 120–132 (1952).

Elliott, J. P., and Flowers, B. H. The structure of the nuclei of mass 18 and 19. *Proc. Roy. Soc.* A229, 536–563 (1953).

Elliott, J. P., Hope, J., and Jahn, H. A. Theoretical studies in nuclear structure. IV. Wave functions for the nuclear p-shell. B. $\langle p^n | p^{n-2} p^2 \rangle$ fractional parentage coefficients. *Phil. Trans. Roy. Soc.* A*246* (*912*), 241–279 (1953).†

Flowers, B. H. Magnetic moments and quadrupole moments of odd-mass nuclei in jj-coupling. *Phil. Mag.* *43*, 1330–1334 (1952).

Jahn, H. A. Theoretical studies in nuclear structure. IV. Nuclear d^2, d^3, and d^4 configurations. Fractional parentage coefficients and central force matrix elements. *Proc. Roy. Soc.* A*205*, 192–237 (1951).†

Jahn, H. A., and van Wieringen, H. Theoretical studies in nuclear structure. IV. Wave functions for the nuclear p-shell. A. $\langle p^n | p^{n-1} p \rangle$ fractional parentage coefficients. *Proc. Roy. Soc.* A*209*, 502–524 (1951).†

Jahn, H. A. Direct evaluation of fractional parentage coefficients using Young operators. General theory and $\langle 4|2, 2\rangle$ coefficients. *Phys. Rev.* *96*, 989–995 (1954).†

Lane, A. M. Studies in intermediate coupling: the energy states of C and ^{13}N belonging to the configuration $1p^9$. *Proc. Phys. Soc.* A*66*, 977–994 (1954).

Lane, A. M., and Radicati, L. A. Studies in intermediate coupling. II: Radiative transitions in light nuclei. *Proc. Phys. Soc.* A*67*, 167–180 (1954).

Lane, A. M., and Wilkinson, D. H. Concept of parentage of nuclear states and its importance in nuclear reaction phenomena. *Phys. Rev.* *97*, 1199–1204 (1955).

Levinson, C., and Ford, K. W. *Phys. Rev.* *99*, 792 (1955).

Meshkov, S. *Phys. Rev.* *91*, 871 (1953).

Racah, G. Theory of complex spectra. III. *Phys. Rev.* *63*, 367 (1943).

Racah, G. Theory of complex spectra. IV. *Phys. Rev.* *76* (*9*), 1352–1365 (1949).

Racah, G. *Lectures on Group Theory.* Institute for Advanced Study, Princeton, 1951.

Redmond, P. J. An explicit formula for the calculation of fractional parentage coefficients. *Proc. Roy. Soc.* A*222*, 84–93 (1954).

Schwarz, C., and de-Shalit, A. Many-particle configurations in a central field. *Phys. Rev.* *94*, 1257–1266 (1954).

(vii) *Nuclear spectra* (see also (vi))

Elliott, J. P. Theoretical studies in nuclear structure. V. The matrix elements of non-central forces with an application to the $2p$ shell. *Proc. Roy. Soc.* A*218*, 345–370 (1953).

Hope, J. The tensor force interaction between a shell closed except for a single vacancy and an external nucleon. *Phys. Rev. 89*, 884 (1953).

de-Shalit, A. The energy levels of odd-odd nuclei. *Phys. Rev. 91*, 1479–1486 (1953).

Talmi, I. The evaluation of the energy matrix of the tensor forces. *Phys. Rev. 89*, 1065–1071 (1953).

(viii) *Nuclear reactions*

Blatt, J. M., and Biedenharn, L. C. The angular distribution of scattering and reaction cross sections. *Revs. Mod. Phys. 24*, 258–272 (1952).

Biedenharn, L. C., and Rose, M. E. Theory of angular correlations of nuclear radiations. *Revs. Mod. Phys. 25*, 729–777 (1953).

Simon, A., and Welton, T. A. Production of polarized particles in nuclear reactions. *Phys. Rev. 90*, 1036–1043 (1953); Errata. *Phys. Rev. 93*, 1435 (1954).

Simon, A. Theory of polarized particles and gamma rays in nuclear reactions. *Phys. Rev. 92*, 1050–1060 (1953).

Tamor, S. Polarization in scattering by complex nuclei. *Phys. Rev. 97*, 1077–1083 (1955).

Wolfenstein, L. Conservation of angular momentum in the statistical theory of nuclear reactions. *Phys. Rev. 82*, 690–696 (1951).

INDEX

Addition of angular momenta, 31-35
Addition theorem, for spherical harmonics, 63
Angles, Eulerian, 6
Angular momentum
  addition of, 31-35
  in classical mechanics, 10
  commutation rules, 10-11
  eigenvalues, 14-18
  eigenvectors, 16, 19-22
  in terms of Euler angles, 13
  generalized, 13-18
  for large l, 27-29
  orbital, 11, 19
  quantization, 10
  of quantum in field represented by vector
  representation of operators of, 13-18
  spherical harmonic, 83
  spin, 17, 25-27
  of system of particles, 12-13
  total, 11
Approximations for large j-values
  for matrix elements of tensor operators, 76-77
  for vector-coupling coefficients and 6-j symbols, 122-123

Basis of representation, 8, 9
  for angular momentum, 14-16
Biedenharn-Elliott sum rule, 96-97
Bohr, quantization of angular momentum, 10

Casimir's convention for Eulerian angles, 6
Clebsch-Gordan coefficients, see vector-coupling coefficients
Coefficients, vector-coupling, see vector-coupling coefficients
Coester and Jauch's coefficient, 105
Commutation relations
  in angular momentum addition, 35
  for components of angular momentum, 11, 14
  for position and linear momentum, 10-11
Condon and Shortley phase convention
  for angular momentum eigenfunctions, 21, 87
  for vector-coupling coefficients, 37
Constants of motion, 4
Contact transformation, infinitesimal, 3

Contragredient transformation, under rotations, 25, 45-46
Coulomb excitation of nuclei, 123
Coupling of angular momenta, 31-35

Degree of representation, 8
Differential operator expressions
  for angular momentum components, 11
  in terms of Eulerian angles, 13
  in spin space, 26-27
Dirac, displacement operators, 11
Double-bar matrix elements, 75
$\mathfrak{D}^{(j)}{}_{m'm}(\alpha\beta\gamma)$ (matrix elements of finite rotations), 55-67
  table of, 129

Eckart, see Wigner-Eckart theorem
Eigenfunctions of orbital angular momentum, 19-25
  of symmetric top, 65-6
Eigenvectors of angular momentum, 14, 19-27
Electric quadrupole moment of proton or electron, 78-79
Elliott-Biedenharn sum rule, 96-98
Emission and Absorption of particles, 85-86
Equivalent representation, 8
Euler angles 6-8
Euler geometrical equations, 66

Factorials, table of powers of prime factors of, 128
Faithful representation, 8
Fano's X-function, 105
Figure axis (in rotation of rigid body), 7

Gradient formula, 79-80
  and vector spherical harmonics, 84
Group
  abstract, definition of, 5
  full linear, $GL(n)$, 5
  of Hamiltonian, 4
  order of, 5
  orthogonal, $O(n)$, 6
  orthogonal unimodular, $SO(3)$, 6
  rotation, 6
  theory, elementary, 5
  unitary, $U(n)$, 6
  unitary unimodular, $SU(2)$, 27

Hamiltonian
  group of, 4
  symmetry of, 3

INDEX

Harmonic polynomials 68-69
  table of, 124
Heisenberg S-matrix, 85
Hermitian adjoint of tensor operators, 77-78
Hermitian product, 14
Hermitian representation for angular momentum operators, 13
Hyperfine structure
  of atom or molecule, 72-73
  of symmetric top molecule, 115-117
  magnetic, 118-119
Hyperfine transitions, intensities, 119

Infinitesimal rotations, 4
  and angular momentum operators, 11
Integrals
  involving $\mathfrak{D}^{(j)}{}_{m'm}(\alpha\beta\gamma)$, 62
  involving $Y_{lm}(\theta, \varphi)$, 63
Interaction
  central 2-body, 113-115
  Gaussian, 114
  Yukawa, 114
Invariants, construction from vector-coupling coefficients, 90-108
Invariant subspace, 9

Jacobi polynomials, 57-58
  and matrix elements of finite rotations, 58-59
Jahn's U-coefficient, 97

Laplace equation, harmonic polynomials as solutions of, 69
Legendre functions, associated, 22-24
  and angular momentum eigenfunctions, 24
Legendre polynomials, 22-23
Line of nodes (in rotation of rigid body), 7
LS coupling, sum rule for transitions, 120
LS-jj coupling transformation coefficients, 107-108

Measurement of angular momentum, 18-19
Metric tensor, for construction of invariants, 46, 93
Momentum angular, see angular momentum
Multipole fields, 84

Nine-j symbol, 100-107
  definition, 101
  evaluation, 105-107
  orthogonality and sum rules, 103-104
  symmetries, 102-103

Normalization
  of harmonic polynomials, 69
  of orbital angular momentum eigenfunctions, 20-21

Operators
  of angular momentum, 11
  tensor (see also tensor operators), 71-73
  vector, 72
Orbital angular momentum, 11, 13, 19-22, 29
Order, of group, 5
Orthogonal groups $O(n)$, $O(3)$, $SO(3)$, 6
Orthogonal transformations and reduction of tensors, 68
Orthogonality
  of angular momentum eigenvectors, 14
  of associated Legendre functions, 24
  of Legendre polynomials, 23
  of vector-coupling coefficients, 38
  of 3-j symbols, 47
  of 6-j symbols, 96
  of 9-j symbols, 103

Parity
  of spherical tensor operators, 72
  of vector spherical harmonics, 83, 84
Pauli spin matrices, 17-18
Phases
  choice for angular momentum eigenfunctions, 21
  choice for eigenvectors of coupled system, 36-37
  of certain vector-coupling coefficients, 39-40
Plane wave, expansion of in spherical harmonics, 80-81
Polar coordinates, relation to Euler angles, 7
Polarization of spherical tensors, 71, 124
Polynomials
  harmonic, 68-69
  Jacobi, 57-58
  Legendre, 22-23
Precession of angular momentum vectors, 32
Probability density
  in classical model of angular momentum addition, 31-32
  of particle in circular orbit, 28-29
Product representation, reduction of, 35, 74

Quadrupole moment, electric, of electron or proton, 78-79
Quadrupole splitting, in hyperfine structure of molecule, 115-117

INDEX

Quantization of angular momentum, 10, 18-19

Racah
  V-coefficient, 52
  W-coefficient, 97
Reality
  of matrix elements of reaction matrix, 86-87
  of vector-coupling coefficients, 40
Recoupling coefficients, for three angular momenta, 91
Recursion relations
  for matrix elements of finite rotations, 61
  for vector-coupling coefficients, 38-39
  for 3-j symbols, 48-49
  for 6-j symbols, 98-99
Reducibility of group representation, 9
Representations
  basis of, 8
  equivalent, 8
  faithful, 8
  irreducible, 4, 9, 17
  products of, 35, 74
  theory of, 8-9
Rigid body, rotation of, 6-8
Rotations, finite, 6-8
  and angular momentum eigenvectors, 53-59
  convention about positive sense of, 6
  matrix elements of, 55-66
    as angular momentum eigenfunctions, 64-65
    computation of, 61-62
    and Jacobi polynomials, 57-59
    as symmetric top eigenfunctions, 65-66
    symmetries of, 59-60
Rotations, infinitesimal, 4

Schwinger
  S-function, 105
  X-coefficient, 52
Selection rules for matrix elements in coupled system, 35-36
Six-j symbol
  definition, 92
  evaluation, 97-99
  orthogonality and sum rules, 96
  symmetries, 94
  tables, 130-132
  tabulations, references to, 99-100
  in terms of 3-j symbols, 92-93
Slater integrals, generalized, 115
S-matrix, 85
Sommerfeld, quantization of direction of angular momentum, 10

Spherical harmonics
  addition theorem for, 63
  as angular momentum eigenfunctions, 22
  example of irreducible representation, 9
  matrix elements of, in jj-coupling, 113
  table of, 124
  vector (see vector spherical harmonics)
Spherical unit vectors, 82
Spin
  differential operators, 26-27
  eigenvectors, 25-27
  matrices, 17-18
  representation of angular momentum, 17
  spherical harmonics, 85
Subgroup, definition of, 5
Sum rules
  for LS coupling transitions, 120
  for 6-j symbols, 96
  for 9-j symbols, 103
Symbols (see also three-j symbols, etc.)
  3-j, 45-50
  6-j, 92-100
  9-j, 100-107
  12-j, 107
Symmetric top, 65-66
  molecule, hyperfine structure of, 115-117
Symmetry
  of Hamiltonian, 3
  of matrix elements of finite rotations, 59-60
  of orbital angular momentum eigenfunctions, 21
  of nine-j symbol, 102-103
  rotational, 4
  of six-j symbol, 94
  of three-j symbol, 46-47
  of vector-coupling coefficient, 41-42

Tensor interaction between nucleons, 73
Tensor operators, 71-73
  factorization of matrix elements of, 73-75
  Hermitian adjoint of, 77-78
  matrix elements of scalar product of, 111
  matrix elements of tensor product of, 109-111
  notations for, 87-88
  reduced matrix elements of, 75-76
    computation of, 76
  scalar product of, 72, 111
  self-adjoint, 78
Tensor product of tensor operators, 69-70, 72, 73, 109-111

INDEX

Tensors, Cartesian, 68
  symmetric and antisymmetric, 68
Tensors, spherical, 68-71
Three-j symbols, 45-50
  computation of symbols with $m_1 = m_2 = m_3 = 0$, 49-51
  recursion relations, 48
Time reversal
  and angular momentum operators, 29-30
  and eigenvectors resulting from vector-coupling, 51-52, 85
  and elements of reaction matrix, 86-87
  and vector-coupling coefficients, 41
Top, symmetric, 65-66
Top, asymmetric, 67
Transformation, infinitesimal, 3
Transition probabilities
  for absorption and emission of particles, 85-86
  in hyperfine structure, 119
  and reduced matrix elements, 76
  sum rule in LS coupling, 120
Twelve-j symbols, 107

U-coefficient of Jahn, 97
Uncertainty principle and quantization of angular momentum, 18-19, 29

Unimodular group, 6
Unitary group, 6

Vector-coupling coefficients, 37-42
  comparison of notations, 52
  computation of, 42-45
  phases of, 39-40
  reality of, 40
  symmetry of, 41-42
  tables, 125-127
  tabulation, references to, 50
Vector spherical harmonics, 81-85
Vertical (in rotation of a rigid body), 7

W-coefficient of Racah, 97, 99
Wigner coefficient, see vector-coupling coefficient, three-j symbol.
Wigner's convention for Euler angles, 6
Wigner-Eckart theorem, 4
Wigner, time reversal and spin, 30
WKB method, applied to angular momentum eigenfunctions, 27-29

Z-coefficient of Biedenharn, Blatt and Rose, 97, 99
Zeeman effect, computation of, 112-113